Introduction to Combinatorics

INTRODUCTION TO COMBINATORICS

Gerald Berman and K. D. Fryer

UNIVERSITY OF WATERLOO

ACADEMIC PRESS   NEW YORK and LONDON

ACADEMIC PRESS, INC.
111 Fifth Avenue, New York, New York 10003

United Kingdom Edition published by
ACADEMIC PRESS, INC. (LONDON) LTD.
24/28 Oval Road, London NW1 7DD

LIBRARY OF CONGRESS CATALOG CARD NUMBER: 70-182646

AMS (MOS) 1970 Subject Classification: 05-01

PRINTED IN THE UNITED STATES OF AMERICA

Contents

Preface   ix
Acknowledgments   xiii

1. Introductory Examples

1.1   A Simple Enumeration Problem   1
1.2   Regions of a Plane   8
1.3   Counting Labeled Trees   13
1.4   Chromatic Polynomials   17
1.5   Counting Hairs   22
1.6   Evaluating Polynomials   23
1.7   A Random Walk   27

Part I.   ENUMERATION

2. Permutations and Combinations

2.1   Permutations   35
2.2   r-Arrangements   38
2.3   Combinations   42
2.4   The Binomial Theorem   45
2.5   The Binomial Coefficients   48
2.6   The Multinomial Theorem   56
2.7   Stirling's Formula   58

3.   The Inclusion–Exclusion Principle

3.1   A Calculus of Sets   60
3.2   The Inclusion–Exclusion Principle   64
3.3   Some Applications of the Inclusion–Exclusion Principle   67
3.4   Derangements   70

4.   Linear Equations with Unit Coefficients

4.1   Solutions Bounded Below   73
4.2   Solutions Bounded Above and Below   78
4.3   Combinations with Repetitions   82

5.   Recurrence Relations

5.1   Recurrence Relations   84
5.2   Solution by Iteration   87
5.3   Difference Methods   90
5.4   A Fibonacci Sequence   94
5.5   A Summation Method   96
5.6   Chromatic Polynomials   96

6.   Generating Functions

6.1   Some Simple Examples   109
6.2   The Solution of Difference Equations by Means of Generating
        Functions   112
6.3   Some Combinatorial Identities   116
6.4   Additional Examples   118
6.5   Derivatives and Differential Equations   121

Part II.   EXISTENCE

7.   Some Methods of Proof

7.1   Existence by Construction   129
7.2   The Method of Exhaustion   132
7.3   The Dirichlet Drawer Principle   136
7.4   The Method of Contradiction   138

8. Geometry of the Plane

8.1	Convex Sets	140
8.2	Tiling a Rectangle	142
8.3	Tessellations of the Plane	147
8.4	Some Equivalence Classes	152

9. Maps on a Sphere

9.1	Euler's Formula	156
9.2	Regular Maps in the Plane	161
9.3	Platonic Solids	163

10. Coloring Problems

10.1	The Four Color Problem	164
10.2	Coloring Graphs	167
10.3	More about Chromatic Polynomials	168
10.4	Chromatic Triangles	174
10.5	Sperner's Lemma	175

11. Finite Structures

11.1	Finite Fields	180
11.2	The Fano Plane	186
11.3	Coordinate Geometry	189
11.4	Projective Configurations	193

Part III.   APPLICATIONS

12. Probability

12.1	Combinatorial Probability	200
12.2	Ultimate Sets	204

13. Ramifications of the Binomial Theorem

13.1	Arithmetic Power Series	208
13.2	The Binomial Distribution	212
13.3	Distribution of Objects into Boxes	214

13.4   *Stirling Numbers*   216
13.5   *Gaussian Binomial Coefficients*   218

14.   *More Generating Functions and Difference Equations*

14.1   *The Partition of Integers*   226
14.2   *Triangulation of Convex Polygons*   230
14.3   *Random Walks*   233
14.4   *A Class of Difference Equations*   239

15.   *Fibonacci Sequences*

15.1   *Representations of Fibonacci Sequences*   242
15.2   *Diagonal Sums of the Pascal Triangle*   245
15.3   *Sequences of Plus and Minus Signs*   246
15.4   *Counting Hares*   249
15.5   *Maximum or Minimum of a Unimodal Function*   250

16.   *Arrangements*

16.1   *Systems of Distinct Representatives*   258
16.2   *Latin Squares*   260
16.3   *The Kirkman Schoolgirl Problem*   263
16.4   *Balanced Incomplete Block Designs*   265
16.5   *Difference Sets*   268
16.6   *Magic Squares*   271
16.7   *Room Squares*   273

Answers to Selected Exercises   277

Index   293

Preface

Combinatorics, or discrete mathematics, and its applications are becoming increasingly important. Polya has said that Combinatorics is an experimental science today just as analysis was decades ago. It is well that students encounter this branch of mathematics at an early level so that they may appreciate that Combinatorics has become a partner with traditional mathematics and with computer science. This book is written to provide an introductory course at the sophomore or junior level.

Because there is so much elementary Combinatorics it is not necessary to wait until the senior years to study the subject. Extensive prerequisites are not necessary. Some knowledge of permutations and combinations, mathematical induction, the binomial theorem, and set theory will allow the student to investigate a host of combinatorial problems and applications. Matrices and determinants are useful but are not prerequisites at the level of this book.

It is possible to select topics which present to the student some quite challenging mathematics. Indeed, in offering him a kaleidoscope of interesting and easily understood topics, chosen to appeal to his imagination, it is often possible to point to some current related research. In addition, a beginning course in these areas can have considerable charm. Its charm does not come from a lack of discipline in the course but from the mathematics involved.

Much Combinatorics has arisen from games and puzzles. Giants such as Gauss, Euler, and Hamilton were interested in puzzles and J. L. Synge has said "The mind is at its best when at play." It is not inappropriate to exploit this point of view in an introductory combinatorics course, again, one hopes stimulating an increase in interest in Mathematics on the part of the student.

A formal definition of Combinatorics is difficult to formulate. Combinatorial problems occur in every branch of mathematics. Roughly speaking, Combinatorics is a study of the arrangements of elements into sets and deals with two general types of problems, enumeration and existence. Recent activity in Combinatorics has been stimulated by applications to other subjects. Thus it seemed logical to us to organize this book into three sections, Enumeration, Existence, and Applications. These three sections follow a chapter of introductory examples. Each of these sections has its own introduction which the reader may consult for further information.

The authors have provided more material than normally can be covered in a term course so that the instructor using the book will have considerable latitude in selecting his course content.

The book is primarily problem-oriented. Exercises appear at the end of each section. We believe that mathematics can be learned only by *doing* mathematics and this involves the solution of a wide range of problems. In offering a course using this text we have found that formal lectures are not always necessary. We have experimented successfully with the group method. The class has been divided into groups of five students, each group with a leader from among the five students. The groups have spent classroom hours primarily discussing theory and working problems, with supplementary lectures from time to time, where deemed necessary. Graduate students have been available both in the classroom and outside at specified tutorial hours to help group leaders prepare for their next class discussion or to clear up unsolved problems. It is our hope that the students gain more understanding from this active involvement in the class structure.

In order that this book be readable and, we hope, of interest to and useable by a wide range of students, we have written with an approach somewhere between the intuitive and the rigorous, but much closer to the former. We have discussed theorems, for example, in a number of different ways. Some we have proved formally; more have been presented informally. Occasionally proofs are demonstrated by examples which give all the steps and reasoning necessary for a formal proof and the reader is asked to complete many of these in the Exercises. Some difficult theorems are stated without proof but with references given.

Combinatorics has become an important tool of the computer scientist. For this reason we have attempted to provide problems that could be of interest to the student of computer science. These problems can be omitted by students who do not have access to a computer without loss of comprehension in the rest of the course.

A course based on this book can be useful not only to the student of computer science and to the mathematics major but also to the student in liberal arts or social science, especially in economics and psychology where Combinatorics is beginning to play an important role.

Finally, we hope that this book will be to many students what the title states, an *introduction* to Combinatorics, which will lead him to a desire to learn more about this fascinating subject.

Acknowledgments

We wish to express our appreciation to:

Ross Honsberger for his contribution to the original version of this book;

E. Anderson, R. N. Burns, R. G. Dunkley, C. E. Haff, W. I. Miller, T. Jenkyns, and P. Schellenberg for reading the manuscript and making valuable suggestions;

Students Bill and Pat Cunningham, Burt Hartnell, Joe Horton, Bill Richardson, and Scott Vanstone for their assistance with the manuscript, particularly with the problems and answers;

Mrs. G. Murison and Mrs. E. Parks for their patience in typing the manuscript many, many times;

Joan Bennewies for her work on the index;

Professors R. C. Mullin, C. St. J. A. Nash-Williams, W. T. Tutte of Waterloo, F. Harary of Michigan, G. Forsythe of Stanford, and Gian-Carlo Rota of Massachusetts Institute of Technology for their kind words and encouragement;

The Faculty of Mathematics of the University of Waterloo for making the computer available for preparing the index (a combinatorial problem!).

Introductory Examples

This chapter contains a number of simple examples that illustrate basic ideas of combinatorics. Important techniques such as the use of recurrence relations and topics such as graph theory are introduced in simplified form. The concepts presented here will be used and expanded on throughout the remainder of the book. Indeed, the reader will find in this chapter the beginnings of a number of threads that are woven through the book.

1.1 A Simple Enumeration Problem

How many nonempty subsets does a finite set have? The answer to this question can be found in a number of ways. Two of these are presented to illustrate the use of transformations, techniques such as induction, and concepts such as recurrence relations, difference equations, and generating functions. It is then shown that the solution to this problem is equivalent mathematically to the solutions of two entirely different problems related to the tower of Hanoi puzzle and the chessboard–grains of wheat caper.

Let s_n denote the number of nonempty subsets of a set S of cardinality n, that is, a set containing n elements. Clearly $s_1 = 1$, since $\{a\}$ has only one nonempty subset, namely the set itself. Also $s_2 = 3$ since $\{a, b\}$ has three non-empty subsets, $\{a\}$, $\{b\}$, and $\{a, b\}$. Note that $s_1 = 1 = 2^1 - 1$, and that $s_2 = 3 = 2^2 - 1$; then check that $s_3 = 7 = 2^3 - 1$. This suggests that s_n might be equal to $2^n - 1$.

THEOREM   Let s_n denote the number of nonempty subsets of a set S of finite cardinality n. Then

$$s_n = 2^n - 1. \tag{1}$$

Proof (by induction)   Let R be the subset of the set N of positive integers for which the theorem is true. We have seen above that $1 \in R$ since $s_1 = 1 = 2^1 - 1$. Now assume that $k \in R$, that is, assume that $s_k = 2^k - 1$ is the number of nonempty subsets of a set S of cardinality k. If we form a new set S^+ by adding a new element b to S, then S^+ has cardinality $k + 1$. Also S^+ contains the subset $\{b\}$; and, for every nonempty subset A of S, the set S^+ has *two* subsets, A and $A \cup \{b\}$. Thus

$$s_{k+1} = 2s_k + 1, \tag{2}$$

so that

$$\begin{aligned} s_{k+1} &= 2(2^k - 1) + 1 \\ &= 2^{k+1} - 1, \end{aligned}$$

and so $k + 1 \in R$ whenever $k \in R$. Since (i) $1 \in R$, and (ii) $k \in R$ implies $k + 1 \in R$, we conclude from the axiom of mathematical induction that $R = N$, that is, that the theorem is true for all positive integers. This completes the proof of the theorem.

We have proved this theorem in detail using the method of induction which should be well known to the reader. Future applications of this method of proof will not usually be given in such detail.

COROLLARY   The number of subsets of a set S of finite cardinality n is 2^n.

The corollary follows immediately from the theorem since the empty set is considered to be a subset of S.

Equation (2) is an example of a **recurrence relation**, an expression which relates a number in a sequence with one or more previous numbers of that sequence. In this case, the kth term s_k of the sequence is the number of nonempty subsets of a set of cardinality k. Thus Equation (2) tells us that the number of nonempty subsets of a set of cardinality $k + 1$, that is, the $(k + 1)$st term of the sequence, is one plus twice the number of nonempty subsets of a set of cardinality k, that is, one plus twice the kth term. Recurrence relations are studied in Chapter 5. An equation such as (2) is sometimes called a **difference equation**. The terms recurrence relation and difference equation are often used interchangeably.

If we observe that $s_0 = 0$, that is, the number of nonempty subsets of the null set is 0, then, using Equation (2), we can calculate successively $s_1 = 2(0) + 1 = 1$, $s_2 = 2(1) + 1 = 3$, and so on. For any k which is sufficiently small, we can calculate s_k by hand in k steps.

This is clearly a method that can be used to find s_k for much larger values of k with the aid of a computer. The drawback to this method is that it gives no hint as to a formula for s_k in terms of k. Indeed, in some complicated enumeration problems it may happen that a recurrence relation like (2) is known, but it is impossible or impractical to find a formula corresponding to (1).

In this particular example we know that $s_{k+1} = 2s_k + 1$ and that $s_0 = 0$, and we wish to find a formula for s_k. We first make the problem simpler by means of a transformation. We let

$$t_k = s_k + 1,$$

so that, from (2),

$$t_{k+1} - 1 = 2(t_k - 1) + 1,$$

or

$$t_{k+1} = 2t_k \qquad \text{with} \quad t_0 = s_0 + 1 = 1.$$

We will now find a formula for t_k which satisfies this recurrence relation and the new initial condition. The solution to the original problem can then be obtained by employing the inverse of the transformation, namely $s_k = t_k - 1$.

Recall that, from the binomial theorem or from formal long division, we may write

$$\frac{1}{1-r} = 1 + r + r^2 + r^3 + \cdots + r^n + \cdots. \tag{3}$$

We regard r as a formal mark rather than a variable so that the right side of (3) is a *formal power series* in r; we are not concerned about questions of convergence of the infinite series involved.

Now form a formal power series in x in which the coefficients are the terms of the sequence $\{t_n\}$ defined by $t_{k+1} = 2t_k$ with $t_0 = 1$; that is, form

$$f(x) = t_0 + t_1 x + t_2 x^2 + \cdots + t_k x^k + t_{k+1} x^{k+1} + \cdots. \tag{4}$$

We refer to $f(x)$ as the generating function of the sequence $\{t_n\}$. As yet we do not know the value of t_k for $k > 0$. Multiply both sides of (4) by $2x$ and formally subtract corresponding members as shown:

$$
\begin{array}{llll}
f(x) = & t_0 + t_1 x + & t_2 x^2 + \cdots + & t_{k+1} x^{k+1} + \cdots \\
2xf(x) = & 2t_0 x + & 2t_1 x^2 + \cdots + & 2t_k x^{k+1} + \cdots
\end{array}
$$

$$(1 - 2x)f(x) = t_0 + (t_1 - 2t_0)x + (t_2 - 2t_1)x^2 + \cdots + (t_{k+1} - 2t_k)x^{k+1} + \cdots .$$

Note that all coefficients of the form $t_{k+1} - 2t_k$ are zero since $t_{k+1} = 2t_k$ for all k. Thus,

$$(1 - 2x)f(x) = t_0 = 1,$$

and

$$f(x) = \frac{1}{1 - 2x}$$

$$= 1 + 2x + 2^2 x^2 + \cdots + 2^k x^k + \cdots \tag{5}$$

from (3) with $r = 2x$. Comparing coefficients of x from this expression with those of Equation (4) we find that $t_k = 2^k$, and conclude that $s_k = 2^k - 1$, as we have proved previously using induction.

The use of generating functions is considered at greater length in Chapter 6.

TOWER OF HANOI (END OF THE WORLD)

The tower of Hanoi puzzle is said to have appeared in 1883. The puzzle consisted of three pegs with eight circular rings of tapering sizes placed in order (the largest on the bottom) on one of the pegs as shown in Figure 1.1.

Figure 1.1

These rings could be moved one at a time from one peg to another with the proviso that a ring never be placed on top of a smaller-sized ring. The object of the game was to move all the rings to one of the other two pegs, that is, to transfer the "tower" of rings to another peg. We consider the problem of finding the minimim number of moves required to transfer a tower of n rings.

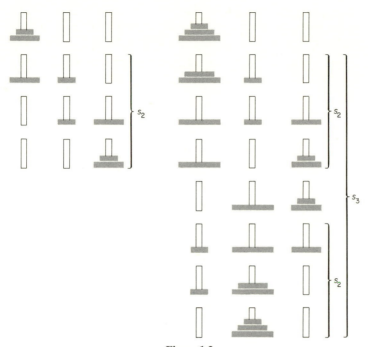

Figure 1.2

Let s_n denote the minimum number of moves required to transfer a tower
of n rings under the condition laid down. Then $s_1 = 1$ and the following
diagrams indicate that $s_2 \leq 3$ and $s_3 \leq 7$ (Figure 1.2). Indeed $s_2 = 3$ and
$s_3 = 7$. Note that in finding s_3 we transfer a tower of two rings twice and a
single ring once so that $s_3 = 2s_2 + 1$. Recalling the problem of finding the
number of nonempty subsets of a finite set we guess that $s_n = 2^n - 1$, and we
may prove this by induction. We sketch only the essential parts of the proof.

The theorem is true for $n = 1$ since $s_1 = 1 = 2^1 - 1$. Now assume that the
theorem is true for $n = k$, that is, $s_k = 2^k - 1$, so that we are assuming a
minimum of $2^k - 1$ moves to transfer a tower of k rings. Now consider a
tower of $k + 1$ rings (Figure 1.3).

Using our induction hypothesis twice, we require $2^k - 1$ moves to produce
the arrangement in Figure 1.4, one move to produce the arrangement in

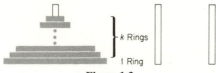

Figure 1.3

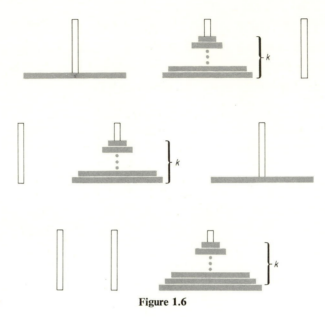

Figure 1.6

Figure 1.5, and $2^k - 1$ moves to produce the final arrangement in Figure 1.6. Hence

$$
\begin{aligned}
s_{k+1} &= (2^k - 1) + 1 + (2^k - 1) \\
&= 2 \cdot 2^k - 1 \\
&= 2^{k+1} - 1,
\end{aligned}
$$

so that the result is true for $n = k + 1$ whenever it is true for $n = k$. The formal details of the induction proof can be filled in by the reader if he feels they are necessary.

Note that to transfer a tower of $k + 1$ rings requires the transfer of a tower of k rings twice and the transfer of a single ring so that $s_{k+1} = 2s_k + 1$. This is the recurrence relation (2) found in the problem of finding the number of nonempty subsets of a finite set.

W. W. Rouse Ball[1] relates the fanciful story that, at the creation, God established a tower of Hanoi arrangement at the temple of Benares using 64 gold disks which are being transferred by priests with the prediction that when all of the disks have been transferred from one diamond needle to another, "with a thunderclap the world will vanish." In Problem 5 of Exercise 1.1 you are asked to check on how much time we have left.

[1] W. W. Rouse Ball, "Mathematical Recreations and Essays," rev. H.S.M. Coxeter, pp. 303–305, Macmillan, London, 1967.

THE CHESSBOARD–GRAINS OF WHEAT CAPER

In his book, "One, Two, Three...Infinity," George Gamow[2] relates the story of King Shirham of India who wished to reward his Grand Vizier, Sissa Ben Dahir, for inventing the game of chess. The vizier spurned great wealth in the form of money or jewels or possessions and made the seemingly modest request of one grain of wheat for the first square on the chessboard, two for the second, four for the third, etc., the number of grains being doubled for each successive square until the final sixty-fourth square was reached. The King was delighted at this paltry request of a few bushels of wheat until his financial advisers worked out the total amount of the payment requested. In Problem 6 of Exercise 1.1 you are asked to estimate the number of bushels of wheat he requested.

Let s_k be the total number of grains of wheat required for the first k squares of the chessboard. Then

$$s_1 = 1 = 2^1 - 1,$$
$$s_2 = 1 + 2 = 3 = 2^2 - 1,$$
$$s_3 = 1 + 2 + 4 = 7 = 2^3 - 1,$$

so again it would appear that $s_k = 2^k - 1$. This can be proved by induction. Assuming $s_k = 2^k - 1$, then 2^k grains are added on the $(k + 1)$st square and

$$s_{k+1} = (2^k - 1) + 2^k$$
$$= 2^{k+1} - 1,$$

so that the result is true for $n = k + 1$ whenever it is true for $n = k$.

Once again the s_k satisfy the recurrence relation (2) which occurs in the problem of finding the number of nonempty subsets of a finite set. This can be shown as follows:

$$s_k = 1 + 2 + 2^2 + 2^3 + \cdots + 2^{k-1}.$$
$$s_{k+1} = 1 + 2 + 2^2 + 2^3 + \cdots + 2^{k-1} + 2^k$$
$$= 1 + 2[1 + 2 + 2^2 + \cdots + 2^{k-1}]$$
$$= 1 + 2s_k.$$

We observe that the three problems of this section are equivalent mathematically. Once the recurrence relation $s_{k+1} = 2s_k + 1$ together with the initial condition $s_1 = 1$ has been established, the solution $s_k = 2^k - 1$ can be obtained by means of any appropriate mathematical technique without making reference to any details of the particular problem.

[2] G. Gamow, "One, Two, Three...Infinity," pp. 19–20, Mentor Books, New American Library, New York, 1954.

EXERCISE 1.1

1. Verify that the number of nonempty subsets of $\{a, b, c, d\}$ is 15 by listing them.

2. Verify that if four rings are used in the tower of Hanoi puzzle, then 15 moves are required.

3. Use induction to prove that the number of subsets (including the empty set) of a set of cardinality n is 2^n. Note that in this case $s_{k+1} = 2s_k$.

4. Apply the generating function method to the recurrence relation (2) without using the transformation first.

    Hint: Let

    $$g(x) = s_0 + s_1x + s_2 x^2 + \cdots + s_k x^k + \cdots .$$

    Multiply both members by $2x$ and formally subtract to obtain

    $$(1 - 2x)g(x) = x + x^2 + x^3 + \cdots$$

    $$= \frac{x}{1 - x},$$

    so that

    $$g(x) = \frac{1}{1 - 2x} - \frac{1}{1 - x}.$$

    Now find s_k, the coefficient of x^k in $g(x)$.

5. You may check, if you wish, that

    $$2^{64} - 1 = 18,446,744,073,709,551,615.$$

    Assume that the priests in the temple of Benares work day and night unceasingly, making one move every second. Find the approximate number of billions of years required to transfer the 64 golden disks in their tower of Hanoi puzzle.

6. Assuming that a bushel of wheat contains 5,000,000 grains, estimate the number of bushels of wheat required to satisfy Sissa Ben Dahir's request in the chessboard–grains of wheat caper.

1.2 Regions of a Plane

A line separates a plane into two regions. Two intersecting lines divide a plane into four regions. In this section we consider the problem of determining the number of regions into which a plane is separated by n lines in general position, that is, with no two lines parallel and no three lines concurrent (passing through a common point). Two methods are presented for determining this number, the method of induction and a method usually studied in a subject called finite differences.

Let $f(n)$ denote the number of regions into which the plane is divided by n lines in general position. We have seen that $f(1) = 2$ and $f(2) = 4$. The careless student might jump to the conclusion that $f(3) = 6$ or $f(3) = 8$. However, a simple diagram shows us that $f(3) = 7$.

In Figure 1.8 a second line l_2 has been added to the first line l_1 of Figure 1.7. This line l_2 intersects l_1 and is thus divided into two rays. Each of these rays divides an existing region into two regions, so that l_2 adds two more regions, and so $f(2) = f(1) + 2$. Now in Figure 1.9 a third line l_3 is added. This line meets each of l_1 and l_2 in distinct points (no two lines are parallel and no

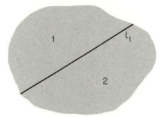

Figure 1.7

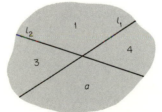

Figure 1.8

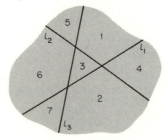

Figure 1.9

three are concurrent). Thus l_3 is divided into three segments, and each of these divides an existing region into two parts (region 1 is divided into 1 and 5, region 3 into 3 and 6, region 2 into 2 and 7); that is, l_3 adds three new regions so that $f(3) = f(2) + 3$.

Suppose a fourth line l_4 is added in such a way that no two of the lines are parallel and no three are concurrent. Then l_4 must intersect l_1, l_2, and l_3 in distinct points and is divided into four segments each of which divides an existing region into two regions. Thus $f(4) = f(3) + 4$.

In general we will have the recurrence relation

$$f(k + 1) = f(k) + (k + 1). \tag{1}$$

To see this, consider the plane divided into $f(k)$ regions by k lines in general position. If line l_{k+1} is added so that no two lines are parallel and no three concurrent as in Figure 1.10, then l_{k+1} meets lines l_1, l_2, ..., l_k in k distinct

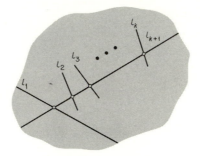

Figure 1.10

points and so is divided into $k + 1$ segments, each of which divides an existing region into two regions. This means that $k + 1$ new regions are created, and so

$$f(k + 1) = f(k) + (k + 1)$$

as required.

We now try to find a formula for $f(n)$ in terms of n when $f(n)$ obeys the recurrence relation (1) together with the initial condition $f(1) = 2$. One possible procedure is indicated in Table 1.1. It would appear that

$$\frac{f(n) - 1}{n} = \frac{n + 1}{2} \qquad \text{or} \qquad f(n) = 1 + \frac{n(n + 1)}{2}.$$

We conjecture that this is the required formula and prove it by induction.

TABLE 1.1

n	$f(n)$	$f(n) - 1$	$\dfrac{f(n) - 1}{n}$
1	2	1	$1 = \frac{2}{2}$
2	4	3	$\frac{3}{2}$
3	7	6	$2 = \frac{4}{2}$
4	11	10	$\frac{5}{2}$
5	16	15	$3 = \frac{6}{2}$

THEOREM   Let $f(n)$ denote the number of regions into which a plane is divided by n lines, no two of which are parallel and no three concurrent; then

$$f(n) = 1 + \frac{n(n + 1)}{2}. \qquad (2)$$

Proof (by induction)   Let S be the subset of positive integers n for which the theorem is true. Then $1 \in S$ since

$$f(1) = 2 \qquad \text{and} \qquad \left[1 + \frac{n(n + 1)}{2} \right]_{n=1} = 2.$$

(The symbol used above means that n is replaced by 1 in the formula.) Now assume $k \in S$, that is, assume that

$$f(k) = 1 + \frac{k(k + 1)}{2}.$$

From the recurrence relation (1),

$$f(k + 1) = f(k) + (k + 1)$$

$$= 1 + \frac{k(k + 1)}{2} + (k + 1)$$

$$= 1 + \frac{(k + 1)(k + 2)}{2}$$

$$= \left[1 + \frac{n(n + 1)}{2} \right]_{n=k+1}.$$

Thus (i) $1 \in S$ and (ii) $k \in S$ implies $k + 1 \in S$, so that $S = N$ by the axiom of induction; and the theorem is true for all positive integers.

Instead of using induction directly we may proceed as follows. Write down Equation (1) for $k = 1, 2, \ldots, n$ and add corresponding members.

$$f(2) = f(1) + 2$$
$$f(3) = f(2) + 3$$
$$\vdots$$
$$f(n) = f(n - 1) + n$$

$$f(2) + f(3) + \cdots + f(n) = f(1) + f(2) + \cdots + f(n - 1) + (2 + 3 + \cdots + n).$$

Canceling the common terms yields

$$f(n) = f(1) + (2 + 3 + \cdots + n)$$
$$= 1 + (1 + 2 + 3 + \cdots + n) \qquad [\text{since } f(1) = 2]$$
$$= 1 + \frac{n(n + 1)}{2}$$

using the fact that the sum of the first n natural numbers is $\frac{1}{2}n(n + 1)$. As we mentioned at the beginning of this section, this is a method studied in finite differences. It has the advantage that the formula need not be known in advance as is the case with inductive proofs.

EXERCISE 1.2

1. Show by means of a diagram that $f(4) = 11$.

2. Into how many regions will the plane be divided by (a) three lines, (b) four lines, (c) five lines, if no two lines are parallel and exactly three are concurrent?

3. If n lines in a plane are such that no two are parallel but exactly three are concurrent, into how many regions will the plane be separated?

4. If three lines in a plane are such that exactly two are parallel, into how many regions will the plane be separated?

5. If four lines in a plane are such that exactly two are parallel and no three concurrent, into how many regions is the plane separated?

6. Repeat Problem 5 for five lines with exactly three parallel but no three concurrent.

7. If $n + k$ lines in the plane are such that exactly k lines are parallel and no three concurrent, into how many regions will the plane be divided?

1.3 Counting Labeled Trees

A simple geometrical object that has many applications is a tree. A tree may be constructed in the ordinary plane as follows.

(i)   Start with an arbitrary point P_1.
(ii) Add a new point P_2 joined by a line to P_1.
(iii) Add a new point P_3 joined by a line to *one* of P_1 or P_2.
(iv) Add a new point P_{i+1} joined by a line to *one* of $P_1, P_2, \cdots, P_i$.

The points P, P_1, P_2, etc. involved are called the **points** or **vertices** or **nodes** of the tree and the line segments or curves involved are the **lines** or **branches** or **edges** of the tree. Each edge has exactly two end points, but a vertex or node may be the end point of several edges. The construction of a tree is indicated in Figure 1.11.

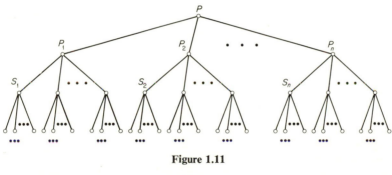

Figure 1.11

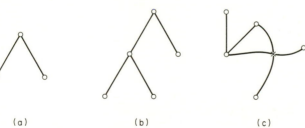

(a)   (b)   (c)

Figure 1.12

Some simple examples of trees are shown in Figure 1.12. Any intersections of the edges in the plane other than the end points of the edges are not considered as vertices of the tree. Thus the reader can see that the trees in (b) and (c) of Figure 1.12 have the same number (five) of vertices and the same number (four) of edges and cannot be distinguished mathematically. The intersection P of the two curves in (c), although it is a point of the plane, is not considered a point or vertex of the tree.

There is essentially only one tree with three vertices, namely the tree in (a) of Figure 1.12. If the vertices of a tree are labeled with different symbols, the tree is called a **labeled tree**. For example, if we use the symbols A, B, and C to label the tree with three vertices, there are three possible labelings, as shown in Figure 1.13. The labeling obtained by interchanging B and C in the

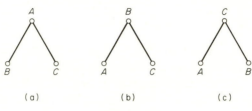

(a)   (b)   (c)

Figure 1.13

tree (a) is not distinguishable from the original labeling except insofar as the drawing is made in the plane. The property that A be joined to both B and C is still true, and the labeled tree is the same. In these trees, one of the vertices is distinguishable from the other two because it is the end point of two edges, whereas the other two are end points of only one edge each. Thus when this special vertex is labeled, it does not matter how we label the others. The labelings in Figure 1.14 are not essentially different, but the labelings in Figure 1.15 are different. The reader should convince himself of the truth of these statements.

Trees were first studied by Cayley in the nineteenth century in an attempt to enumerate chemical compounds. Using his results he was able to predict

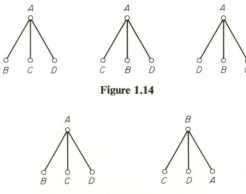

Figure 1.14

Figure 1.15

the existence of several organic compounds which were subsequently isolated.

It is a difficult problem to count the number of trees with a given number of vertices. However, Cayley found an elegant way to count labeled trees. His method uses the multinomial theorem (the binomial theorem generalized to more than two terms) which we shall encounter in Chapter 2 and which plays an important role in combinatorics.

In our demonstration of Cayley's method we will restrict our attention to trees with four vertices. Before illustrating Cayley's approach we will enumerate all the labeled trees with four vertices directly by labeling the two essentially different trees with four vertices shown in Figure 1.16. (It is easily verified that these are the only essentially different trees with four vertices.) In

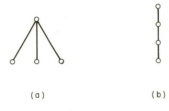

(a)   (b)

Figure 1.16

Figure 1.16a, one of the vertices has **degree** three, that is, it is joined to three other vertices and the others have degree one; in Figure 1.16b, two of the vertices have degree two and two have degree one.

We will use the symbols x_1, x_2, x_3, and x_4 to label the trees, as Cayley did in his method. In Figure 1.16a, once we have labeled the special vertex of degree three (which can be done in four ways, choosing x_1, x_2, x_3, or x_4), it does not matter how we label the other vertices. Hence there are four labelings of this particular tree.

For the tree in Figure 1.16b, we may label one of the vertices of degree two in four ways and then label the second vertex of degree two in three ways; it then does not matter how the remaining two vertices of degree one are labeled. Thus there are 12 labelings of this tree, and so altogether there are 16 labelings of a tree with four vertices.

In Figure 1.17 we show the 16 labelings of a tree with four vertices. The unlabeled vertices can be labeled in any way without producing more than one essentially different labeled tree in each case.

It should be clear to the reader that this method will be difficult to carry out, if not impractical or indeed impossible, if we wished to find the number of labeled trees with, say, 17 vertices.

Cayley used the following approach. In a labeled tree, associate the product $x_i x_j$ with the edge joining the vertices x_i and x_j and with the tree

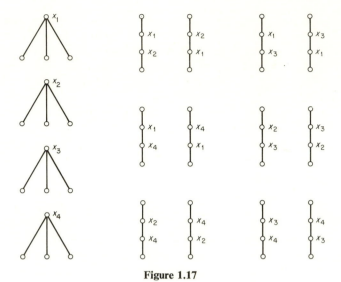

Figure 1.17

associate the product of all such products. Thus for the labeled tree in Figure 1.18a we associate the product

$$(x_1 x_2)(x_1 x_3)(x_1 x_4) \qquad \text{or} \qquad x_1{}^3 x_2 x_3 x_4.$$

The reader should check that there is a one-to-one correspondence between the labelings of this tree and expressions of this form by writing down the

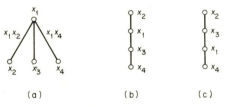

(a)   (b)   (c)

Figure 1.18

labeled trees corresponding to the products $x_1 x_2{}^3 x_3 x_4$, $x_1 x_2 x_3{}^3 x_4$, and $x_1 x_2 x_3 x_4{}^3$. The labeled trees in Figures 1.18b and 1.18c both correspond to the product

$$(x_2 x_1)(x_1 x_3)(x_3 x_4) \qquad \text{or} \qquad x_1{}^2 x_2 x_3{}^2 x_4.$$

The sum S of all products connected with labeled trees with four vertices is

$$S = \sum x_1{}^3 x_2 x_3 x_4 + 2 \sum x_1{}^2 x_2 x_3{}^2 x_4$$

where the summation is over all products of the given form. This expression can be simplified.

$$
\begin{aligned}
S &= x_1 x_2 x_3 x_4 \left(\sum x_1^2 + 2 \sum x_1 x_3 \right) \\
&= x_1 x_2 x_3 x_4 (x_1^2 + x_2^2 + x_3^2 + x_4^2 \\
&\quad + 2x_1 x_2 + 2x_1 x_3 + 2x_1 x_4 + 2x_2 x_3 + 2x_2 x_4 + 2x_3 x_4) \\
&= x_1 x_2 x_3 x_4 (x_1 + x_2 + x_3 + x_4)^2.
\end{aligned}
\tag{1}
$$

Each term in the expansion of the right-hand side of (1) corresponds to a labeled tree, and each labeled tree corresponds to a term in the expansion. The number of labeled trees is thus the number of terms in the expansion, and this can be obtained by setting each variable x_i equal to 1; that is,

$$
[S]_{x_i = 1} = 1(1 + 1 + 1 + 1)^2 = 4^2,
$$

so that there are 4^2 or 16 labeled trees as we have already verified.

In general, the labeled trees with n vertices correspond to the terms in the expansion of

$$
S = x_1 x_2 \cdots x_n (x_1 + x_2 + \cdots + x_n)^{n-2}
$$

so that the number of labeled trees on n vertices is n^{n-2}. We will not prove this result in general. The expansion of the factor $(x_1 + x_2 + \cdots + x_n)^{n-2}$ in S is an application of the multinomial theorem.

EXERCISE 1.3

1. Use induction to prove that a tree with n vertices or nodes has $n - 1$ edges or branches.

2. (a) Show that there are three trees with five vertices.
    (b) Find the number of (distinct) labelings of each of the trees in (a) and hence show that there are 125 labeled trees with five vertices.

3. (a) Verify that the labeled trees in (2) correspond to the terms of the expansion of

$$
x_1 x_2 x_3 x_4 x_5 (x_1 + x_2 + x_3 + x_4 + x_5)^3.
$$

    (b) Deduce from (a) that there are 125 labeled trees with five vertices.

1.4 Chromatic Polynomials

In this section we will be concerned with coloring ordinary maps, that is, maps which are drawn on a plane surface (a piece of paper) or on a sphere (globe). Such maps consist of a collection of regions (countries) separated by boundaries or edges. (We assume that no country is made up of two or more disjoint parts.) In coloring a map, different colors must be used for neighboring regions, that is, for regions having a nontrivial common boundary (more than a single point). G. D. Birkhoff showed in 1912 that the number of ways, $P(M, \lambda)$, of coloring a map M with λ colors can be expressed as a polynomial in λ,

$$P(M, \lambda) = C_0 + C_1\lambda + C_2\lambda^2 + \cdots + C_k\lambda^k$$

for some choice of integers $C_0, C_1, \ldots, C_k$ where k is the number of regions. This polynomial is called the **chromatic polynomial** of the map.

Consider the simple maps M_1, M_2, M_3, and M_4 in Figure 1.19, and suppose we wish to color these maps using λ colors. For the map M_1, we may

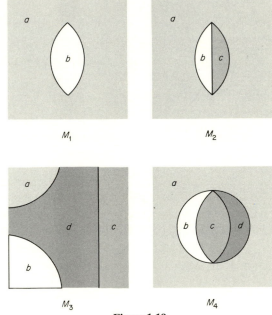

Figure 1.19

use any one of the λ colors for the region a and any one of the remaining $\lambda - 1$ colors for region b so that

$$P(M_1, \lambda) = \lambda(\lambda - 1) = \lambda^2 - \lambda.$$

For the map M_2, there will be λ choices of colors for the region a, then $\lambda - 1$ for region b and $\lambda - 2$ for region c, so that

$$P(M_2, \lambda) = \lambda(\lambda - 1)(\lambda - 2) = \lambda^3 - 3\lambda^2 + 2\lambda.$$

For M_3, there will be λ choices of colors for region d, and any one of the remaining colors can be used for regions a, b, and c, so that

$$P(M_3, \lambda) = \lambda(\lambda - 1)^3 = \lambda^4 - 3\lambda^3 + 3\lambda^2 - \lambda.$$

The coloring of the map M_4 is slightly more complicated; regions a and b must be colored differently, but a and c can be the same color. Region a can be colored in λ ways, and then regions b and d can be colored the same or differently, and so we must consider two cases.

Case 1. Regions b and d are the same color. Then there are λ choices for region a, $\lambda - 1$ choices for the common color for b and d, and $\lambda - 1$ choices for c (it can be the same color as a). Thus there are $\lambda(\lambda - 1)^2$ different colorings with b and d colored the same.

Case 2. Regions b and d are colored differently. There are λ choices for region a, $\lambda - 1$ for region b, $\lambda - 2$ for region d, and so $\lambda - 2$ for region c (it cannot be the same as b or d). Thus there are $\lambda(\lambda - 1)(\lambda - 2)^2$ colorings in this case.

Combining the results of the two cases we have

$$P(M_4, \lambda) = \lambda(\lambda - 1)^2 + \lambda(\lambda - 1)(\lambda - 2)^2$$
$$= \lambda^4 - 4\lambda^3 + 6\lambda^2 - 3\lambda.$$

As asserted in Birkhoff's theorem, in each of these examples the expression for the number of ways of coloring the given maps is a polynomial in which the degree is the number of regions in the corresponding map. In particular, note that if three colors are used, that is, if $\lambda = 3$, then M_1 can be colored in 6 ways, M_2 in 6 ways, M_3 in 24 ways, and M_4 in 18 ways.

When we speak of coloring a map, for example M_3, using four colors, we include the possibility of using fewer colors. If we had the colors red, blue, green, and orange, we could use all four colors or we could color d red and a, b, and c green (or blue or orange) or we could color d red, a and b green, and c blue (or orange) etc. If we were asked for the number of ways of coloring M_3 using *exactly* four colors (never fewer), the answer would be 4! or 24, the number of ways of permuting the four colors. Note that $P(M_3, 4) = 4 \cdot 3^3 = 108$ and that this includes colorings using red, blue, green, and orange, but not necessarily all of them in each coloring. Let $n(M, \lambda)$ be the number of ways of coloring the map M using *exactly* λ colors. Then $n(M_3, 4) = 24$, as we have seen above. Also $n(M, 1) = 0$ for any map consisting of more than one region.

Finally, we prove Birkhoff's theorem.

THEOREM   Let M be a map of k regions in which no region is made up of two or more disjoint parts and let $P(M, \lambda)$ be the number of ways of coloring M with λ colors so that adjacent regions have different colors. Then $P(M, \lambda)$ is a polynomial of degree k.

Proof   There are (see Section 2.3)

$$\binom{\lambda}{j} = \frac{\lambda(\lambda - 1)(\lambda - 2) \cdots (\lambda - j + 1)}{j(j - 1)(j - 2) \cdots 1}$$

ways of choosing j colors from λ colors and hence $n(M, j) \cdot \binom{\lambda}{j}$ ways of coloring M with exactly j of the λ colors. Then

$$P(M, \lambda) = n(M, 1) \binom{\lambda}{1} + n(M, 2)\binom{\lambda}{2} + \cdots + n(M, \lambda)\binom{\lambda}{\lambda},$$

which is clearly a polynomial in λ. Further, $n(M, k + j) = 0$ for $j > 0$ since it is impossible to use more colors than regions. This proves the theorem.

In Section 1.3 we considered geometrical objects called trees, which are special examples of geometrical objects called **graphs**. A graph consists of vertices and edges with each edge connecting precisely two vertices, as in the definition of a tree. In the case of a graph, *any* pair of vertices can be joined by an edge. Indeed, a vertex can be joined to itself by an edge (called a loop).

A labeled graph can be associated with a labeled map by associating a vertex or node with each region, and then joining two vertices in the graph when the corresponding regions have a common boundary. In Figure 1.20 we show the labeled graphs associated with the labeled maps of Figure 1.19. The labeled graph (a) corresponds to the labeled map M_1, (b) to M_2, (c) to M_3, and (d) to M_4.

Note that the construction of a tree precludes the existence of **circuits** or **closed paths** in the graph. In a tree it is impossible to start at a vertex, move along a set of edges exactly once, and return to the starting point. This is possible in the graphs (b) and (d) of Figure 1.20; these graphs are not trees, although the graphs (a) and (c) are trees.

The reader should convince himself that the coloring of a map so that adjacent regions have different colors is equivalent to the coloring of the vertices of the corresponding graph in such a way that vertices joined by an

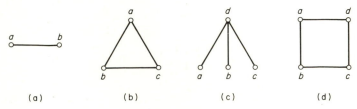

(a)   (b)   (c)   (d)

Figure 1.20

edge have different colors. We define the **chromatic polynomial of a graph** to be the polynomial that expresses the number of ways of coloring the vertices of the graph in such a way that the vertices joined by an edge have different colors. We shall consider chromatic polynomials at greater length in Chapters 5 and 10.

It is a well-known result in graph theory, not proved here, that although every map in the plane corresponds to a graph, it is not true that every graph corresponds to a map in the plane. The reader may consult some elementary work on graph theory for the proof that the graph in Figure 1.21 does not correspond to a map of the type we have been considering, that is, a map that can be drawn on a plane surface or on a sphere with no country consisting of more than one region.

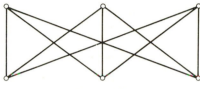

Figure 1.21

EXERCISE 1.4

1. Enumerate the ways of coloring the maps M_1, M_2, M_3, and M_4 of Figure 1.19 using three colors, red, green, and blue.

2. Find the numbers $n(M, 2)$, $n(M, 3)$, and $n(M, 4)$ for the maps M_1, M_2, M_3, and M_4, and use these numbers to deduce the chromatic polynomials of these maps.

3. Represent the map in Figure 1.22 by means of a labeled graph and find the corresponding chromatic polynomial.

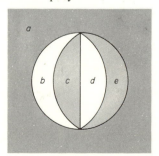

Figure 1.22

4. Show that the chromatic polynomial of a tree with n vertices is $\lambda(\lambda - 1)^{n-1}$.

1.5 Counting Hairs

Does life exist on Mars? This is a question that has been asked many times since the planet was first discovered. Recent evidence seems to indicate that intelligent (?) life as we know it cannot exist. However, what about other types of life? It has been suggested that it might be possible for life to be based on silicon compounds rather than carbon compounds.

To be convinced of the existence of life on Mars we would insist on being shown a specimen or assured that some reputable person had seen a specimen. It would be impossible to have an airtight argument for nonexistence, since we know so little about the universe in which we live. (The reader might like to find out what evidence for the existence or nonexistence of life on Mars and Venus has been discovered.)

Questions concerning the existence or nonexistence of the Abominable Snowman and the Loch Ness monster are similarly awkward ones. However, the situation regarding existence is different in mathematics, especially in dealing with finite collections of objects. One way of showing existence or nonexistence is by enumeration. For example, in Section 1.3 we showed the existence of trees and labeled trees with four vertices. In fact we showed that there are exactly two trees and exactly 16 labeled trees with four vertices by drawing the former and enumerating the latter. As a corollary we can state without reservation that there do not exist three different trees with four vertices or a tree with four vertices such that one of the vertices has degree four, that is, is joined to four other vertices.

Proving existence by enumeration is a direct method. There are also indirect methods of proving existence or nonexistence. The following theorem illustrates the pigeonhole principle or Dirichlet drawer principle, which is discussed further in Chapter 7.

THEOREM There exist two people living in New York City who have the same number of hairs on their heads.

To prove this theorem we assume that we have made up some rules for deciding whether or not a given hair belongs to an individual's head-hair count. Now assign to each individual i a number $h(i)$, the number of hairs on his head. Suppose also that a number H is known such that $h(i) \leq H$, that is, no individual has more than H hairs on his head. Now suppose that we write the values of $h(i)$ for all individuals i on slips of paper, and suppose that there are I slips of paper. Construct a set of pigeonholes labeled $0, 1, 2, \ldots, H$, and put the slips of paper in the pigeonholes corresponding to numbers written thereon. Consider the three cases: $I < H + 1$, $I = H + 1$, and $I > H + 1$. If

$I < H + 1$, we can be sure that at least one of the pigeonholes is empty. If $I = H + 1$, then either there is one slip of paper in each pigeonhole or there is at least one empty pigeonhole and at least one pigeonhole with two slips. If $I > H + 1$, it is apparent that at least one pigeonhole must contain at least two slips.

The theorem will thus follow from the above remarks if the population of New York City is greater than the maximum head-hair count of any individual. Checking this information is left as an exercise for the reader.

Notice that we have no way of knowing which two (or more) individuals have the same head-hair count without a head-hair count census.

EXERCISE 1.5

1. Prove that there does not exist a tree with five vertices with a vertex of degree five.

2. Prove that there does not exist a tree with n vertices with a vertex of degree n.

3. Prove that there exists a tree with n vertices with a vertex of degree m where $m < n$.

4. Consult a reference work to determine an upper bound on the number of human hairs an individual can have on his head. Use the result to prove that there are at least three individuals in New York City with the same head-hair count.

1.6 Evaluating Polynomials

Not too many years ago most arithmetical computations were done by hand, that is, with pencil, paper, and eraser. Today small mechanical or electronic calculators are commonplace, and many businesses, industries, and universities have computing facilities. However, contrary to a popular myth prevalent even fairly recently, computers can do no more than ordinary arithmetic operations, addition, subtraction, multiplication, and division. The difference between computers and humans is that computers can carry out these operations on a much larger scale. Whereas an individual may get tired after making a few hundred computations, or a few thousand on a hand calculator in several minutes, there are now computers that can do a billion calculations in a second, and most of the large computers can do a million or so calculations a second.

If we are required to multiply two three-digit numbers we would probably do the job by hand. If we had twenty such problems to do we would probably seek out a small desk calculator. If we had 10,000 such problems, we would probably want to learn how to use a computer.

Various numerical methods have been developed, and are still being developed, for converting computations to additions and multiplications. These methods can be used for hand computation or when using computing aids for complicated problems. A few ideas along these lines are illustrated in this section by considering the problem of evaluating the chromatic polynomial

$$f(x) = x^4 - 4x^3 + 6x^2 - 3x \tag{1}$$

we encountered in Section 1.4 for various values of x.

If we require, say, only the value $f(3)$, then the form found in the derivation,

$$f(x) = x(x-1)^2 + x(x-1)(x-2)^2$$

would be convenient since we obtain

$$f(3) = 3 \cdot 2^2 + 3 \cdot 2 \cdot 1^2 = 18$$

without having to evaluate 3^4.

We could also use the well-known remainder theorem.

REMAINDER THEOREM   If f is a polynomial, then the remainder when $f(x)$ is divided by $x - a$ is $f(a)$.

In our example, the remainder $f(3)$ could be found by dividing $f(x)$ by $x - 3$ using ordinary division or possibly by using the method of synthetic division illustrated in Table 1.2. The first row is a listing of the coefficients of f,

TABLE 1.2

1	−4	6	−3	0	3
	3	−3	9	18	
1	−1	3	6	18	

including any zero coefficients, and to the right is the number a in the divisor $x - a$. The first entry in the third row is the first entry in the first row. The entries in the second row are obtained by multiplying elements in the third row by 3 (in general, by a), and the entries in the third row are the sums of the corresponding entries in the first two rows. The entries are obtained successively. The last entry in the third row gives the remainder, 18, and the preceding entries give the coefficients of the quotient obtained when $f(x)$ is divided by $x - 3$. Thus

$$f(x) = (x - 3)(x^3 - x^2 + 3x + 6) + 18.$$

The reader who is unfamiliar with this method should divide $f(x)$ by $x - 3$ in the usual way to check this result. He should also consult an elementary algebra text for a derivation of this method of synthetic division.

This is a good method for hand computation of small values of polynomials and can easily be programmed for a computer. However, it is simpler to program the polynomial directly, using FORTRAN. If a computer is used to evaluate a polynomial, it is recommended that the terms of the polynomial be bracketed in a special way. In our example we would rewrite $f(x)$ as

$$f(x) = x(-3 + x(6 + x(-4 + x))). \tag{2}$$

This requires the computer to perform three multiplications and three additions whereas the form (1) requires six multiplications and three additions.

If several values of a polynomial are required, any of the above methods can be used to obtain them. However, if it is required to make a table of values of $f(x)$ for successive integral values of x, say $x = -1, 0, 1, 2, \ldots$, then an alternative procedure can be used. This is illustrated in Table 1.3 for the polynomial

$$f(x) = x^4 - 4x^3 + 6x^2 - 3x.$$

TABLE 1.3

x	$f(x)$	$\Delta f(x)$	$\Delta^2 f(x)$	$\Delta^3 f(x)$	$\Delta^4 f(x)$
−1	14	−14	14	−12	24
0	0	0	2	12	(24)
1	0	2	14	(36)	
2	2	16	(50)		
3	18	(66)			
4	(84)				

In the first two columns we write x and $f(x)$ respectively for $x = -1, 0, 1, 2$, and 3. Now define

$$\Delta f(x) = f(x + 1) - f(x) \tag{3}$$

so that $\Delta f(-1)$, $\Delta f(0)$, $\Delta f(1)$, and $\Delta f(2)$ can be obtained from the values $f(-1), f(0), f(1), f(2)$, and $f(3)$. For example,

$$\Delta f(2) = f(3) - f(2) = 18 - 2 = 16.$$

Similarly, we define

$$\Delta^2 f(x) = \Delta(\Delta f(x)) = \Delta f(x + 1) - \Delta f(x);$$

for example,

$$\Delta^2 f(1) = \Delta f(2) - \Delta f(1) = 16 - 2 = 14.$$

Continuing, we define

$$\Delta^3 f(x) = \Delta^2 f(x + 1) - \Delta^2 f(x) \qquad \text{and} \qquad \Delta^4 f(x) = \Delta^3 f(x + 1) - \Delta^3 f(x).$$

Making the corresponding evaluations for the table, we finally obtain $\Delta^4 f(-1) = 24$. This procedure is studied in a subject called finite differences and is usually included in courses in numerical methods. It can be shown in these courses that for this particular polynomial, $\Delta^4 f(x) = 24$ for all values of x. Knowing this, we can work backward, obtaining successively,

$$\Delta^4 f(0) = 24,$$
$$\Delta^3 f(1) = \Delta^3 f(0) + \Delta^4 f(0) = 36,$$
$$\Delta^2 f(2) = \Delta^2 f(1) + \Delta^3 f(1) = 50,$$
$$\Delta f(3) = \Delta f(2) + \Delta^2 f(2) = 66,$$
$$f(4) = f(3) + \Delta f(3) = 84,$$

the circled entries in Table 1.3. In this way we have obtained a new value of the polynomial, namely $f(4)$. Continuing in this way we can obtain $f(5)$, $f(6)$, etc.

From now on a letter **c** will be used in the Exercises to indicate problems which are best handled with the aid of a calculator or computer because of the number of computations involved. The reader who must omit these problems because of the unavailability of a calculator or computer will in no way be at a disadvantage in pursuing the work of this text.

EXERCISE 1.6

1. (a) Use the remainder theorem and synthetic division to evaluate $f(5)$, $f(6)$, $f(7)$, and $f(8)$ for the following polynomials:

   (i) $f(x) = x^4 - 4x^3 + 6x^2 - 3x$
   (ii) $f(x) = 3x^4 + x^3 - 2x^2 + x - 6$
   (iii) $f(x) = x^3 - 3x + 2.$

   (b)ᶜ Write a program to carry out long division of a polynomial by a monomial $x - a$. Check the program using the results in part (a) and then evaluate $f(12)$, $f(13)$, and $f(18)$ for each polynomial.

2.[c] Program Equation (2) for the computer and then evaluate $f(12)$, $f(13)$, and $f(18)$.

3. (a) Continue evaluating the missing entries in Table 1.3 to obtain $f(5), f(6), f(7)$, and $f(8)$ for the polynomial

$$f(x) = x^4 - 4x^3 + 6x^2 - 3x.$$

   (b) Set up similar tables for the polynomials in (ii) and (iii) of 1(a) and use these tables to evaluate $f(5), f(6), f(7)$, and $f(8)$ for these polynomials.

   (c)[e] Program the method of Table 1.3 and evaluate $f(x)$ for $x = 5, 6, \ldots,$ 100 for the three polynomials of Problem 1.

4. A **monic** polynomial of degree n is one in which the coefficient of x^n is $+1$.

   (a) Show that $\Delta^2 f(x) = 2$ if $f(x)$ is any monic quadratic polynomial in x.
   (b) Show that $\Delta^3 f(x) = 6$ if $f(x)$ is any monic cubic polynomial in x.
   (c) Conjecture and prove a generalization of (a) and (b).

1.7 A Random Walk

In this section we assume that the reader understands the elementary notions of probability. The idea of a random walk is introduced in a special case that is related to the binomial theorem. Again we discover the usefulness of generating functions and difference equations. Probability is considered in greater detail in Chapter 12 and the binomial theorem in Chapters 2 and 14.

Suppose that a particle P is at the origin of a cartesian coordinate system at time $t = 0$. Every unit of time, say every second, the particle makes a decision to stay at its present position or move one unit to the right along the x axis. We assume that the probability that P does not move in a given unit of time is d and the probability that P moves to the right in a given unit of time is p, where $p + d = 1$.

Let the probability that the particle is at the point $(n, 0)$ at the end of the kth interval of time be denoted by $P^k(n)$. It is easy to verify that

$$P^0(0) = 1, \qquad P^0(n) = 0 \qquad \text{for} \quad n \neq 0;$$
$$P^1(0) = d, \qquad P^1(1) = p, \qquad P^1(n) = 0 \qquad \text{for} \quad n \neq 0, 1;$$
$$P^2(0) = d^2, \qquad P^2(1) = 2pd, \qquad P^2(2) = p^2.$$

A general formula is presented in the following theorem.

THEOREM   A particle starts at the origin at time $t = 0$, and at each interval of time stays in its present position with probability d or moves one unit to the right with probability p, with $p + d = 1$. Let $P^k(n)$ denote the probability that the particle is at the point $(n, 0)$ at the end of the kth interval of time; then

$$P^k(n) = \binom{k}{n} p^n\, d^{k-n},$$

where

$$\binom{k}{n} = \frac{k(k-1)(k-2)\cdots(k-n+1)}{n!}$$

if $n \leq k$, and is equal to zero otherwise.

We have already checked the validity of this theorem above for $n = 0, 1,$ and 2. The theorem can be proved using induction after we establish the following lemma.

LEMMA $P^k(n) = pP^{k-1}(n-1) + dP^{k-1}(n)$.

The lemma follows from the definition of the random walk. The only way that P can be at the point $(n, 0)$ after k intervals of time is to have been at $(n - 1, 0)$ after $k - 1$ intervals of time and move to the right, which it does with probability p, or to have been at the point $(n, 0)$ after $k - 1$ intervals of time and stay there, which it does with probability d.

In the proof of the main theorem we require the following result which the reader is asked to verify from the definition of the symbols involved:

$$\binom{k-1}{n-1} + \binom{k-1}{n} = \binom{k}{n}. \tag{1}$$

Now the main part of the induction proof is to assume the theorem is true for $k - 1$, that is, assume

$$P^{k-1}(n) = \binom{k-1}{n} p^n d^{k-1-n} \qquad \text{for all} \quad n,$$

and note that

$$P^k(n) = pP^{k-1}(n-1) + dP^{k-1}(n) \qquad \text{(from the lemma)}$$

$$= p\binom{k-1}{n-1} p^{n-1} d^{k-n} + d\binom{k-1}{n} p^n d^{k-n-1}$$

$$\text{(from the induction hypothesis)}$$

$$= \left[\binom{k-1}{n-1} + \binom{k-1}{n} \right] p^n d^{k-n}$$

$$= \binom{k}{n} p^n d^{k-n} \qquad\qquad \text{[using (1)].}$$

The details of the induction are left to the reader.

COROLLARY $P^k(n)$ is the coefficient of x^n in the binomial expansion of $(px + d)^k$.

Proof   By the binomial theorem,

$$(px + d)^k = \sum_{n=0}^{k} \binom{k}{n}(px)^n d^{k-n}$$

$$= \sum_{n=0}^{k} \left[\binom{k}{n}p^n d^{k-n}\right] x^n$$

$$= \sum_{n=0}^{k} P^k(n)x^n.$$

The polynomial $(px + d)^k$ is called the **generating polynomial** of the probabilities $P^k(n)$. The equation in the lemma is the **difference equation** or **recurrence relation** of the random walk. The conditions $P^0(0) = 1$, and $P^0(n) = 0$ for $n \neq 0$, are the **initial** or **boundary conditions** of the random walk.

EXAMPLE 1   In the random walk described above, what is the probability that the particle P is to the right of the point $(2, 0)$ at the end of (a) three, (b) four intervals of time?

Solution   (a)   probability $= P^3(3) = p^3$ since $P^3(n) = 0$ for $n > 3$;
(b)   probability $= P^4(3) + P^4(4) = 4p^3 d + p^4$.

The point at which the particle is most likely to be at the end of k intervals of time will be the point or points associated with the maximum of the probabilities

$$P^k(0), \ P^k(1), \ \ldots, \ P^k(k).$$

Finally, we define the **mean position** of the particle in the random walk at the end of k intervals of time to be $\bar{x}_k$ where

$$\bar{x}_k = \sum_{n=0}^{k} nP^k(n).$$

EXAMPLE 2   Find the mean position of the particle P in the random walk described above at the end of (a) two, (b) three intervals of time.

Solution   (a) $\bar{x}_2 = 0P^2(0) + 1P^2(1) + 2P^2(2)$
$\qquad = 2pd + 2p^2$
$\qquad = 2p(d + p) = 2p.$
(b) $\bar{x}_3 = 0P^3(0) + 1P^3(1) + 2P^3(2) + 3P^3(3)$
$\qquad = 3pd^2 + 6p^2 d + 3p^3$
$\qquad = 3p(d + p)^2 = 3p.$

EXERCISE 1.7

Problems 1–5 refer to the definitions and the random walk described in the theorem of this section.

1. Using the definitions verify the values of $P^k(n)$ for $k = 3$ and 4.

2. What is the probability that at the end of the fifth interval of time the particle is to the left of the point $(3, 0)$?

3. If $p = \frac{1}{3}$, what is the probability that the particle is at the point (a) $(3, 0)$, (b) $(5, 0)$ at the end of seven intervals of time?

4. If $d = \frac{1}{2}$, at what point(s) will the particle most likely be at the end of four intervals of time?

5. Show that (a) $\bar{x}_4 = 4p$, (b) $\bar{x}_5 = 5p$, (c) $\bar{x}_k = kp$.

6.* Assume that in a random walk the particle starts at the origin at time $t = 0$, and either stays in its present position at any interval of time with probability d, moves one unit to the right with probability p, or moves one unit to the left with probability q, where $p + q + d = 1$. Show that the difference equation of this random walk is

$$P^k(n) = pP^{k-1}(n-1) + qP^{k-1}(n+1) + dP^{k-1}(n)$$

and that $P^k(n)$ is the coefficient of x^n in the expansion of $(px + qx^{-1} + d)^k$.

---

* Problems marked with an asterisk are of a more challenging nature and may be omitted without detracting from the work of this text.

I

ENUMERATION

Several basic methods of enumeration are discussed in the following chapters. These include the use of permutations and combinations in Chapter 2, the inclusion–exclusion principle in Chapter 3, linear equations with unit coefficients in Chapter 4, recurrence in Chapter 5, and generating functions in Chapter 6. Many of the ideas concerning permutations and combinations discussed in Chapter 2 will be already known to the reader. The methods of this chapter are useful in applications. The ideas of recurrence are most important in computations involving the computer, but theoretical arguments frequently employ generating functions.

Permutations and Combinations

In the preface we stated that we are assuming that the reader has some familiarity with permutations and combinations, and so much of the material of this chapter should be review material. The problems in the exercises should serve to reinforce the reader's knowledge of this area.

A number of the derivations in this and in succeeding chapters are based on the following *multiplication principle*.

THE MULTIPLICATION PRINCIPLE

If a set of objects can be *separated* or *partitioned* into k nonempty disjoint subsets, and if each of these subsets can be separated into m nonempty disjoint subsets, then the original set can be separated into km nonempty disjoint subsets.

EXAMPLE 1   In how many ways can a man and a woman be selected from eight men and five women ?

Solution   Consider the set S of ordered pairs (m, w) where m is one of the eight men and w is one of the five women. The set S can be partitioned into eight subsets

$$\{(m_1, w)\}, \{(m_2, w)\}, \ldots, \{(m_8, w)\}$$

where $\{(m_i, w)\}$ is the collection of all pairs in which m_i is the man. Each of these subsets can be separated into five subsets (each containing one element), for example, $\{(m_2, w)\}$ is partitioned into the five subsets

$$\{(m_2, w_1)\}, \{(m_2, w_2)\}, \ldots, \{(m_2, w_5)\}.$$

By the multiplication principle, the original set has now been partitioned into $8 \times 5 = 40$ disjoint subsets, each of which provides an appropriate selection, and there are no other possible selections.

EXAMPLE 2   In how many ways can a man, a woman, and a child be selected from eight men, five women, and seven children ?

Solution   Let S be the set of ordered triples (m, w, c) where c is one of the seven children. Partition S into seven subsets

$$\{(m, w, c_1)\}, \{(m, w, c_2)\}, \ldots, \{(m, w, c_7)\}$$

where $c_1, c_2, \ldots, c_7$ are the seven children. According to Example 1, each of these subsets can be partitioned into 40 subsets containing a different pair (m, w). By the multiplication principle, the number of different triples is $40 \times 7 = 280$ and this is the required number. Notice that $280 = 8 \times 5 \times 7$.

These examples illustrate how the multiplication principle can be extended to a finite number of applications.

The following symbols, in which n is a real number and r a positive integer, are used in this chapter:

(i) $0! = 1$
         $r! = r(r - 1)(r - 2) \cdots 3 \cdot 2 \cdot 1$

(ii) $n^{(0)} = 1$
         $n^{(r)} = n(n - 1)(n - 2) \cdots (n - r + 1)$

(iii) $\displaystyle \binom{n}{r} = \frac{n^{(r)}}{r!} = \frac{n(n - 1)(n - 2) \cdots (n - r + 1)}{r!}$

(iii)′ $\displaystyle \binom{n}{r} = \frac{n!}{r!(n - r)!}$ if n is a positive integer and $n \geq r$.

If $n, r_1, r_2, \ldots, r_k$ are positive integers with $n = r_1 + r_2 + \cdots + r_k$, then

(iv) $\displaystyle \binom{n}{r_1, r_2, \ldots, r_k} = \frac{n!}{r_1! r_2! \cdots r_k!}.$

2.1 Permutations

A **permutation** of a set of objects is a mapping of the set onto itself. The symbol

$$\begin{pmatrix} A & B & C \\ B & A & C \end{pmatrix}$$

represents the permutation of the set $\{A, B, C\}$ in which A is mapped onto B, B is mapped onto A, and C is mapped onto C.

EXAMPLE 1   There are six permutations of the set $\{A, B, C\}$.

$$\begin{pmatrix} A & B & C \\ A & B & C \end{pmatrix}, \quad \begin{pmatrix} A & B & C \\ A & C & B \end{pmatrix}, \quad \begin{pmatrix} A & B & C \\ B & A & C \end{pmatrix},$$

$$\begin{pmatrix} A & B & C \\ B & C & A \end{pmatrix}, \quad \begin{pmatrix} A & B & C \\ C & A & B \end{pmatrix}, \quad \begin{pmatrix} A & B & C \\ C & B & A \end{pmatrix}.$$

The second line in each case is an ordered arrangement of the elements of the first line. The number of permutations of the set $\{A, B, C\}$ is the number of ordered arrangements of the symbols A, B, C. This number is six and the ordered arrangements are

$$ABC, \; ACB, \; BAC, \; BCA, \; CAB, \; CBA.$$

The permutation

$$\begin{pmatrix} A & B & C \\ A & B & C \end{pmatrix}$$

in which each element or symbol is mapped onto itself is called the identity permutation I for this set.

THEOREM   The number of distinct permutations on a set of finite cardinality n is $n!$.

Proof   The number of permutations on the set $\{x_1, x_2, \ldots, x_n\}$ will be the number of ordered arrangements of the symbols $x_1, x_2, \ldots, x_n$. There are n choices for the first element, then $n - 1$ for the second, $n - 2$ for the third, etc. Applying the multiplication principle, the total number is

$$n(n - 1)(n - 2) \cdots 3 \cdot 2 \cdot 1 = n!,$$

proving the theorem.

The *product* $P \circ Q$ of two permutations P and Q on the same set is defined to be the single permutation on the set equivalent to the successive performance of permutations P and Q (P first, Q second). The symbol $\circ$ is used for multiplication instead of $\cdot$ or $\times$ to stress the fact that we are not dealing with ordinary multiplication, which would have no meaning applied to permutations.

EXAMPLE 2   We label the permutations on the set $\{A, B, C\}$ in Example 1, I, P_1, P_2, P_3, P_4, and P_5 respectively, and calculate the products (i) $P_2 \circ P_4$ and (ii) $P_4 \circ P_2$.

(i) $P_2 \circ P_4$ means "carry out mapping P_2 first on $\{A, B, C\}$ and then the mapping P_4 on the resultant arrangement":

$$
\begin{array}{ccc}
 & \text{under } P_2 & \text{under } P_4 \\
A \longrightarrow & B \longrightarrow & A \\
B \longrightarrow & A \longrightarrow & C \\
C \longrightarrow & C \longrightarrow & B.
\end{array}
$$

The single permutation equivalent to this product or composition of mappings is

$$
\begin{pmatrix} A & B & C \\ A & C & B \end{pmatrix} = P_1,
$$

so that $P_2 \circ P_4 = P_1$.

(ii) $P_4 \circ P_2$ means "permutation P_4 followed by permutation P_2"

$$
\begin{array}{ccc}
 & \text{under } P_4 & \text{under } P_2 \\
A \longrightarrow & C \longrightarrow & C \\
B \longrightarrow & A \longrightarrow & B \\
C \longrightarrow & B \longrightarrow & A.
\end{array}
$$

Thus

$$
P_4 \circ P_2 = \begin{pmatrix} A & B & C \\ C & B & A \end{pmatrix} = P_5.
$$

Note that in this example, $P_2 \circ P_4 \neq P_4 \circ P_2$.

EXERCISE 2.1

1.   Complete the multiplication table for the six permutations $I, P_1, \ldots, P_5$ of Example 1.

$\circ$	I	P_1	P_2	P_3	P_4	P_5
I						
P_1						
P_2				P_1		
P_3						
P_4		P_5				
P_5						

2. Find $P \circ Q$ and $Q \circ P$ for the following permutations,

(a) $P = \begin{pmatrix} 1 & 2 & 3 \\ 3 & 2 & 1 \end{pmatrix}$, $\qquad Q = \begin{pmatrix} 1 & 2 & 3 \\ 2 & 3 & 1 \end{pmatrix}$

(b) $P = \begin{pmatrix} 1 & 2 & 3 & 4 \\ 4 & 3 & 1 & 2 \end{pmatrix}$, $\qquad Q = \begin{pmatrix} 1 & 2 & 3 & 4 \\ 2 & 4 & 3 & 1 \end{pmatrix}$

(c) $P = \begin{pmatrix} 1 & 2 & 3 & 4 & 5 \\ 5 & 2 & 1 & 3 & 4 \end{pmatrix}$, $\qquad Q = \begin{pmatrix} 1 & 2 & 3 & 4 & 5 \\ 1 & 5 & 3 & 4 & 2 \end{pmatrix}$

(d) $P = \begin{pmatrix} a & b & c & d & e \\ b & d & c & e & a \end{pmatrix}$, $\qquad Q = \begin{pmatrix} a & b & c & d & e \\ e & d & c & b & a \end{pmatrix}$.

3. If $P^2 = P \circ P$, find P^2 and Q^2 for each permutation in (2).

4. Find $P \circ (Q \circ R)$ and $(P \circ Q) \circ R$ in the following,

(a) P and Q from 2(a), $R = \begin{pmatrix} 1 & 2 & 3 \\ 1 & 3 & 2 \end{pmatrix}$

(b) P and Q from 2(b), $R = \begin{pmatrix} 1 & 2 & 3 & 4 \\ 4 & 1 & 3 & 2 \end{pmatrix}$

(c) P and Q from 2(c), $R = \begin{pmatrix} 1 & 2 & 3 & 4 & 5 \\ 4 & 5 & 3 & 1 & 2 \end{pmatrix}$

(d) P and Q from 2(d), $R = \begin{pmatrix} a & b & c & d & e \\ c & a & b & e & d \end{pmatrix}$.

5. Prove that multiplication of permutations is associative.

6. Prove that multiplication of permutations is, in general, not commutative.

7. If $P \circ Q = I$, where I is the identity permutation, we say that P and Q are inverses and write $Q = P^{-1}$. If

$$P = \begin{pmatrix} 1 & 2 & 3 & 4 \\ 2 & 4 & 1 & 3 \end{pmatrix},$$

then

$$P^{-1} = \begin{pmatrix} 2 & 4 & 1 & 3 \\ 1 & 2 & 3 & 4 \end{pmatrix} = \begin{pmatrix} 1 & 2 & 3 & 4 \\ 3 & 1 & 4 & 2 \end{pmatrix}.$$

Find the inverses of all the permutations P and Q in (2).

8.* Prove that $(P \circ Q)^{-1} = Q^{-1} \circ P^{-1}$.

2.2 *r*-Arrangements

By an **r-arrangement** of n distinct objects we mean an ordered selection of r of the objects. We use the symbol $P(n, r)$ to represent the number of r-arrangements of n distinct objects. The six 3-arrangements of the distinct symbols A, B, and C are

$$ABC, \; ACB, \; BAC, \; BCA, \; CAB, \; CBA.$$

THEOREM 1   Let $P(n, r)$ be the number of r-arrangements of n distinct objects, then

$$P(n, r) = n(n - 1)(n - 2) \cdots (n - r + 1) = n^{(r)}.$$

Proof (1)   Consider r labeled boxes to be filled, each with one object selected from the n distinct objects. There are n choices for the first box, $n - 1$ for the second, ..., $n - r + 1$ for the rth. An application of the multiplication principle completes this proof.

(2)   There are $P(n, r - 1)$ ordered arrangements of n distinct objects taken $r - 1$ at a time. For each such arrangement, there will be $n - r + 1$ remaining objects. Adding one of these objects at the end of an $(r - 1)$-arrangement produces an r-arrangement, and all r-arrangements can be generated in this way. Again applying the multiplication principle we may state that the total number of r-arrangements of n distinct objects is $(n - r + 1)P(n, r - 1)$. Thus

$$P(n, r) = (n - r + 1)P(n, r - 1).$$

By analogy,

$$P(n, r - 1) = (n - r + 2)P(n, r - 2)$$
$$P(n, r - 2) = (n - r + 3)P(n, r - 3)$$
$$\vdots$$
$$P(n, 3) = (n - 2)P(n, 2)$$
$$P(n, 2) = (n - 1)P(n, 1)$$
$$P(n, 1) = n.$$

Now multiply corresponding members of these equations and cancel the common factors to obtain

$$P(n, r) = (n - r + 1)(n - r + 2) \cdots (n - 2)(n - 1)n = n^{(r)},$$

as required.

Note that the number of n-arrangements of n distinct objects is $n^{(n)} = n!$, the number of permutations on a set of n objects.

In considering *r*-arrangements of *n* distinct objects we are making a selection of the objects and of course there can be no repetitions. In some cases we may have duplicates of the *n* distinct objects. Theorem 2 considers such a situation.

THEOREM 2   The number of arrangements of *n* distinct objects taken *r* at a time with unlimited repetitions allowed is n^r.

The reader may prove Theorem 2 using the multiplication principle.

We are sometimes interested in arrangements of distinct objects in a circle. In such a case, only the relative position of the objects is of interest. In Figure 2.1, both diagrams represent the same arrangement. We consider only the relative position of the objects *A*, *B*, *C*, and *D*, and not the positions themselves.

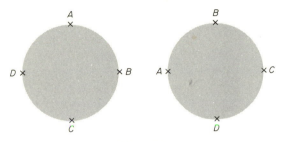

Figure 2.1

THEOREM 3   The number of ordered arrangements of *n* objects in a circle is $(n - 1)!$.

The theorem may be proved by observing that we may arbitrarily seat one person at a round table and we then have $n - 1$ choices for the person to the right of him. For the person to the right of the one just selected we have $n - 2$ choices, and so on. The reader may construct his own proof, applying the multiplication principle.

In Theorem 4, we consider the number of ordered arrangements of a set of *n* objects, not all of which are distinct.

THEOREM 4   The number of *n*-arrangements of *n* objects of which exactly *p* are alike of one kind and exactly *q* are alike of a second kind, with the remainder all different, is

$$\frac{n!}{p!q!}.$$

EXAMPLE   Consider the collection $A = [a, a, a, b, b]$ in which we consider the three a's and two b's to be five distinct objects. Here $n = 5$, $p = 3$, and $q = 2$. According to the formula in Theorem 4, the number of ordered arrangements of all of these objects is

$$\frac{5!}{3!\,2!} = 10.$$

This result is checked by listing the arrangements.

aaabb	aabab	abaab	baaab
aabba	ababa	baaba	
abbaa	babaa		
bbaaa			

Proof of theorem   Let x be the required number of arrangements. Let the n objects be labeled $x_1, x_2, \ldots, x_n$, with the p alike of one kind labeled x_1, $x_2, \ldots, x_p$, and the q alike of the second kind labeled $x_{p+1}, \ldots, x_{p+q}$. Then there will be $n!$ n-arrangements of these (now distinct) labeled objects, and so

$$n! = xp!\,q!$$

since for each of the x required arrangements, the first p alike, being labeled, can be interchanged in $p!$ ways and the q alike of a second kind, being labeled, can be interchanged in $q!$ ways. Thus the theorem is proved.

Finally, Theorem 5 deals with the number of ways of making selections from a given set of objects.

THEOREM 5   The number of ways of making a nonempty selection by choosing some or all of $p_1 + p_2 + \cdots + p_k$ objects where p_1 are alike of one kind, p_2 alike of a second kind, $\ldots$, p_k alike of a kth kind, is

$$(p_1 + 1)(p_2 + 2) \cdots (p_k + 1) - 1.$$

Proof   The p_i like objects yield $p_i + 1$ selections for each i since we can choose either 0, 1, 2, $\ldots$, or p_i objects from that subset. Applying the multiplication principle yields the product $(p_1 + 1)(p_2 + 2) \cdots (p_k + 1)$ from which one selection must be subtracted to ensure that we have at least one element in each of our selections, that is, we must exclude the possibility of taking no element from each of the k subsets.

EXERCISE 2.2

1. How many of the integers between 10,000 and 100,000 have no digits other than 5, 7, or 9? How many have no digits other than 5, 7, 9, or 0?

2. How many different five-digit number license plates can be made if:

    (a)   the first digit cannot be 0 and repetition of digits is not allowed?

    (b)   the first digit cannot be 0 but repetitions are allowed?

3. How many batting orders are possible for a baseball team of nine players if the pitcher always bats in ninth position?

4. In how many ways can a President, Secretary, and Treasurer be chosen from a six-member committee?

5. How many integers between 1000 and 9999 inclusive have distinct digits? Of these, how many are even numbers? How many consist entirely of odd digits?

6. How many integers greater than 64,000 have both the following properties:

    (i)   the digits of the integer are distinct;

    (ii) the digits 3 and 8 do not occur in the number.

7. How many consecutive zeros occur on the right-hand end of the decimal numeral for 30!?

8. In how many ways is it possible to seat six persons at a round table if

    (a)   there are no restrictions on the seating;

    (b)   two of the six persons must not sit in adjacent seats.

9. In how many ways can five men and five ladies be seated at a round table if no two women are to be in adjacent seats ?

10. Starting at a given intersection in a city a man wishes to walk to a second intersection which is eight blocks west and five blocks north. In how many different paths can he walk the thirteen blocks if he never walks east or south ?

11. Find the number of distinct permutations of all the letters of the following words:

    (a)   WATERLOO

    (b)   MATHEMATICS

    (c)   LACKAWANNA

12. How many numbers greater than 4,000,000 can be formed from the digits 2, 3, 4, 4, 5, 5, 5?

13. Two points A and B in space have coordinates $(1, 1, 1)$ and $(3, 7, 10)$, respectively. How many paths lead from A to B if steps of length one unit can be taken only in the positive x, y, or z directions?

14. In how many different orders can the 17 letters

$$x, \ x, \ y, \ y, \ y, \ y, \ y, \ y, \ z, \ z, \ z, \ z, \ z, \ z, \ z, \ z, \ z$$

be written ?

15. How many distinct divisors do the following numbers have:

(a) 3^5 (b) $3^2 5^3$
(c) $3^3 5^2 7^5$ (d) 29,400 ?

16. How many different numbers can be formed by multiplying some or all of the digits

$$2, \ 3, \ 4, \ 4, \ 5, \ 5, \ 5 \ ?$$

2.3 Combinations

A **combination** of a set of objects is an unordered selection of the objects. Repetitions may or may not be permitted.

EXAMPLE 1   There are three combinations of two letters selected from a, b, c if no repetitions are allowed, namely ab, bc, ca, and six combinations with repetitions, adding aa, bb, and cc.

We use $C(n, r)$ as the symbol for the number of combinations of n distinct objects taken r at a time without repetitions. Such combinations are also referred to as **r-subsets** of a set of n elements. Another common symbol for $C(n, r)$ is $\binom{n}{r}$, read "n choose r"; this symbol was used in Section 1.4 and was defined at the beginning of Chapter 2.

THEOREM   The number, $C(n, r)$, of combinations of n distinct objects taken r at a time without repetitions, that is, the number of r-subsets of a set of n elements, is

$$C(n, r) = \frac{n^{(r)}}{r!} = \binom{n}{r}.$$

Proof   There are $r!$ *r*-arrangements of *r* objects. The number of *r*-arrangements of *n* distinct objects must then be $r!\,C(n, r)$ so that

$$r!\,C(n, r) = P(n, r) = n^{(r)}.$$

The theorem follows by division by $r!$

COROLLARY 1 $C(n, r) = \dfrac{n!}{r!\,(n - r)!}.$

COROLLARY 2 $C(n, r) = C(n, n - r).$

EXAMPLE 2

$$C(7, 3) = \binom{7}{3} = \frac{7!}{3!\,4!};$$

$$C(7, 7 - 3) = \binom{7}{4} = \frac{7!}{4!\,3!}.$$

EXERCISE 2.3

1.   A candy store stocks chocolate bars of four different brands. Each brand comes in three different sizes and each size comes in either plain chocolate or chocolate plus nuts. How many different kinds of chocolate bars does the store have?

2.   A shoe store stocks 17 styles of men's shoes. Each style is available in 12 lengths, 4 widths, and 2 colors. How many different shoes does the store stock?

3.   In how many ways can the letters of the word *combine* be arranged if

   (a)   the first letter must be a vowel?
   (b)   the third and sixth letters must be consonants?
   (c)   there are no restrictions?

4.   A student is to answer any ten questions from an examination consisting of fifteen questions. In how many ways can he make his selection?

5.   If the student in Problem 4 is to choose four questions from the first seven and six from the last eight, in how many ways can he make his selection?

 6. Twenty-four points, no three collinear, lie in a plane. How many line segments can be formed having these points as terminal points ? How many triangles can be formed having these points as vertices?

 7. In how many ways can a committee of four be selected from six men and eight women if the committee must contain at least two women and if Mrs. Brown refuses to serve on the same committee as her husband?

 8. In how many ways can a bridge hand of thirteen cards be dealt from a 52-card deck?

 9. In how many ways can α distinct objects be placed in β distinct boxes if there are more boxes than objects and if not more than one object is placed in each box?

10. There are ten steamers plying between New York and Liverpool. In how many ways can a person travel from New York to Liverpool and return by a different steamer?

11. Find the number of ways $m + n$ objects can be divided into two groups containing m and n objects, respectively.

12. Find the number of ways of partitioning fifteen objects into three equal groups.

13. Find the number of ways in which fifteen recruits can be assigned to three different regiments.

14. A man has six friends. In how many ways can he invite one or more to dinner?

15. A music club has sixty members; twenty are sopranos, sixteen are altos, fourteen are tenors, ten are basses.
 (a)   How many quartets (one soprano, one alto, etc.) can be formed in the club?
 (b)   How many double quartets (two sopranos, etc.) can be formed in the club?

16. A crew of eight men is to be chosen to row an eight-oared boat. If there are eleven men, five of whom can row on the stroke side only, four on the bow side only, and the remaining two on either side, in how many ways may the boat be manned?

17.* Prove the following

> THEOREM   The number of combinations of n objects taken r at a time with repetitions allowed is
>
> $$\binom{n+r-1}{n-1} = \binom{n+r-1}{r}.$$

2.4 The Binomial Theorem

We state here the binomial theorem for the expansion of $(x + a)^n$ for n a positive integer. We assume that the reader is familiar with this theorem.

> THEOREM 1   (binomial theorem for n a positive integer)
>
> $$(x + a)^n = x^n + C(n, 1)x^{n-1}a + \cdots + C(n, r)x^{n-r}a^r + \cdots + a^n.$$

Derivation   Consider the product

$$(x + a)(x + a) \cdots (x + a)$$

of n factors $x + a$. In the expansion of this product the term $x^{n-r}a^r$ occurs whenever a is chosen from r of the n factors. Thus there are $C(n, r)$ terms of the form $x^{n-r}a^r$. Since all terms in the expansion will have such a form for some r, the theorem follows when n is a positive integer.

The theorem can be proved using induction after we have derived Pascal's formula in the next section.

The binomial theorem can be extended to real and complex values of n. When n is a nonnegative integer, the expansion of $(x + a)^n$ contains $n + 1$ terms; otherwise, the expansion contains an infinite number of terms. When n is a real number, we use the expression

$$(x + a)^n = x^n + \binom{n}{1}x^{n-1}a + \cdots + \binom{n}{r}x^{n-r}a^r + \cdots.$$

If the terms in the expansion of $(x + a)^n$, for n a positive integer, are written down, then for given x and a, there will be one or more terms with the greatest value. These terms can be predicted by the following theorem.

> THEOREM 2   For fixed n, if $ax > 0$, the maximum term in the expansion of $(x + a)^n$ is the term $C(n, r)x^{n-r}a^r$ for which $r = [\alpha]$, where $[\alpha]$ is the greatest integer $\leq \alpha$, and
>
> $$\alpha = \frac{n + 1}{(x/a) + 1}.$$

Proof   The $(r + 1)$st term in the expansion of $(x + a)^n$ is

$$t_{r+1} = C(n, r)x^{n-r}a^r.$$

The reader should check that

$$\frac{t_{r+1}}{t_r} = \left[\frac{n+1}{r} - 1\right]\frac{a}{x} \tag{1}$$

and that $t_{r+1}/t_r > 1$ provided $\alpha > r$, and for fixed n, the factor $[(n + 1)/r] - 1$ decreases as r increases. Complete the proof.

EXAMPLE 1

$$(2 + 1)^3 = \binom{3}{0}2^3 + \binom{3}{1}2^2 \cdot 1 + \binom{3}{2}2 \cdot 1^2 + 1^3$$
$$= 8 + 12 + 6 + 1.$$

If we apply Theorem 2 to $(2 + 1)^3$ in which $x = 2$, $a = 1$, $n = 3$, we have

$$\alpha = \frac{4}{2+1} = \frac{4}{3}; \qquad r = [\alpha] = \left[\frac{4}{3}\right] = 1.$$

The maximum term then is $\binom{3}{1}2^2 \cdot 1 = 12$, as above.

EXAMPLE 2

$$(3 + 1)^3 = \binom{3}{0}3^3 + \binom{3}{1}3^2 \cdot 1 + \binom{3}{2}3 \cdot 1^2 + 1^3$$
$$= 27 + 27 + 9 + 1.$$

Applying Theorem 2 with $x = 3$, $a = 1$, $n = 3$, we have

$$\alpha = \frac{4}{3+1} = 1; \qquad r = [\alpha] = 1.$$

The maximum term then is $\binom{3}{1}3^2 \cdot 1 = 27$.

Note in Example 2 that Theorem 2 predicts one of the two maximum terms in the expansion of the given binomial. In Problem 10 of Exercise 2.4 you will be asked to extend the statement of Theorem 2 to predict such multiple maximum terms.

EXERCISE 2.4

1.   Consult a standard college algebra text for a proof of the binomial theorem when n is (i) a negative integer, (ii) a rational number, and (iii)* a real number.

2. Find the coefficient of x^5 in the expansion of

$$(x + 2x^{-1})^7.$$

3. What is the coefficient of $a^3 b^6$ in $(a + b)^9$?

4. What is the coefficient of $x^2 y^{11}$ in $(x + y)^{13}$?

5. Show that $(p + q)^7$ can be expressed as the sum of all terms of the form

$$\frac{7!}{a!\,b!} p^a q^b$$

where a and b range over all pairs of nonnegative integers a and b whose sum is seven.

6. Generalize Problem 5 to $(p + q)^n$ where n is a positive integer.

7. Differentiate the series for $(1 + x)^n$ and compare coefficients of x^r on both sides to prove that

$$nC(n - 1, r) = (r + 1)C(n, r + 1).$$

8. Assume that

$$(x + a)^n = c_0 + c_1 x + \cdots + c_n x^n$$

is an identity in x. Set $x = 0$ to obtain $c_0 = a^n$. Differentiate both sides of the identity and obtain c_1. Derive all the coefficients c_i in this way.

9. Check result (1) of this section and show that $t_{r+1}/t_r > 1$ provided $\alpha > r$. Complete the proof of Theorem 2.

10. In Theorem 2, show that the terms

$$C(n, r)x^{n-r} a^r \qquad \text{and} \qquad C(n, r - 1)x^{n-r+1} a^{r-1}$$

are both maximum terms if $r = \alpha$ when

$$\alpha = \frac{n + 1}{(x/a) + 1}$$

is an integer.

11. Find the maximum term(s) in the expansions of

(a) $(5 + 2)^3$ (b) $(9 + 3)^7$ (c) $(12 + 3)^{14}$.

12.   Show that

$$(a + \sqrt{a^2 - 1})^7 + (a - \sqrt{a^2 - 1})^7 = 2a(64a^6 - 112a^4 + 56a^2 - 7).$$

13.   Find the coefficient of x^n in the expansion of

$$\left(x^2 + \frac{1}{x^3}\right)^n.$$

14.   Use the binomial theorem to expand $(x^2 + 2x - 1)^3$.

2.5 The Binomial Coefficients

A great number of interesting and useful results can be obtained involving the binomial coefficients $C(n, r)$. A few of these are discussed in this section. We begin with a recurrence relation known as Pascal's formula which will be used frequently in this book.

PASCAL'S FORMULA   If $C(n, r)$, n a positive integer, is the general binomial coefficient, then

$$C(n, r) = C(n - 1, r) + C(n - 1, r - 1).$$

Derivation $C(n, r)$ is the number of r-subsets of a set of n distinct objects. For a particular object P, the r-subsets each fall into one of two classifications:

(i)   those subsets involving P;
(ii)   those subsets not involving P.

There are $C(n - 1, r - 1)$ r-subsets of type (i) and $C(n - 1, r)$ r-subsets of type (ii). This establishes the formula.

Pascal's theorem can also be proved by using Corollary 1 of the theorem in Section 2.3, that is, by substituting

$$\frac{n!}{r!\,(n - r)!}$$

for $C(n, r)$.

EXAMPLE 1   The set $\{1, 2, 3, 4\}$ has six 2-subsets:

$$\{1, 2\}, \{1, 3\}, \{1, 4\}, \{2, 3\}, \{2, 4\}, \{3, 4\}.$$

The first three contain the element 1 and the remaining three do not. The first three are formed by fixing 1 and choosing one element from {2, 3, 4}; the last three are formed by choosing two elements from {2, 3, 4}. Thus

$$\binom{4}{2} = \binom{3}{1} + \binom{3}{2}$$

which checks with Pascal's formula.

We now introduce a very famous array, known as Pascal's triangle, which consists of the binomial coefficients.

PASCAL'S THEOREM   The binomial coefficients in the expansions of $(x + y)^n$ for $n = 0, 1, 2, \ldots$ may be written in the following array known as Pascal's triangle (Table 2.1). The $(r + 1)$st entry of the $(n + 1)$st row is $C(n, r)$. This table can be constructed and extended by means of the Pascal formula.

TABLE 2.1

```
1
1 1
1 2  1
1 3  3   1
1 4  6   4   1
1 5 10  10   5   1
1 6 15 (20)(15)  6  1
1 7 21 35 (35) 21  7  1
. . .  .   .  . . . . .
```

EXAMPLE 2 $C(7, 4) = C(6, 4) + C(6, 3)$, as circled in the array. This follows from the Pascal formula with $n = 7$, $r = 4$.

Pascal's triangle has many interesting properties. A number of these are referred to in the problems of Exercise 2.5.

The binomial coefficients can be related to the paths in the plane from the origin to points with integral coefficients, as in the following theorem.

THEOREM 1   A particle, starting at the origin, can move one unit at a time either in the positive x direction or the positive y direction is the plane. The number of distinct paths the particle can take from the origin to the point $(n - r, r)$ is $C(n, r)$.

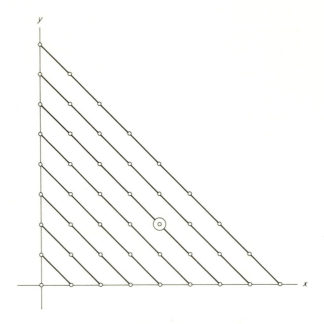

Figure 2.2

EXAMPLE 3 Find the number of distinct paths a particle can take from the origin to the point (4, 2) if the particle can move either one unit in the positive x or y direction at a time (Figure 2.2).

There is but one path to any point $(a, 0)$ on the x axis or to any point $(0, b)$ on the y axis. Also, the point (a, b) can be reached either

(i)   by being at the point $(a - 1, b)$ and moving one unit to the right, or
(ii)   by being at the point $(a, b - 1)$ and moving one unit up, as indicated in Figure 2.3.

Thus the number of paths to point (a, b) is the sum of the number of paths to points $(a - 1, b)$ and $(a, b - 1)$.

All of the lattice points (points with integral coefficients) in the first quadrant lie on a family of parallel lines as shown in Figure 2.2 (these are lines with

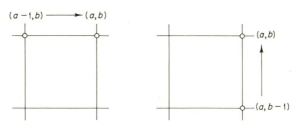

Figure 2.3

equations $x + y = k$ where k is a nonnegative integer). The first line contains the origin; the second, the two lattice points $(0, 1)$ and $(1, 0)$; the third, the three lattice points $(0, 2)$, $(1, 1)$, $(2, 0)$; etc. We may list the number of paths to the lattice points in each member of this family of parallel lines in the following pattern:

$$
\begin{array}{ccccccccccccc}
 & & & & & & 1 & & & & & & \\
 & & & & & 1 & & 1 & & & & & \\
 & & & & 1 & & 2 & & 1 & & & & \\
 & & & 1 & & 3 & & 3 & & 1 & & & \\
 & & 1 & & 4 & & 6 & & 4 & & 1 & & \\
 & 1 & & 5 & & 10 & & 10 & & 5 & & 1 & \\
1 & & 6 & & 15 & & 20 & & 15 & & 6 & & 1
\end{array}
$$

and so on. Each entry is the sum of the two above it to the right and to the left. This follows because of the above result that the number of paths to the point (a, b) is the sum of the number of paths to the points $(a - 1, b)$ and $(a, b - 1)$.

The point $(4, 2)$ is on the seventh of the family of parallel lines. The number of paths to it is then the third entry (from either end) in the seventh row of our scheme, namely 15. Note that the number of paths to the point $(3, 2)$ is 10 (third entry in the sixth row) and the number of paths to the point $(4, 1)$ is five (second entry in the sixth row) and $10 + 5 = 15$.

In general, the point $(n - r, r)$ is on the line $x + y = n$, the $(n + 1)$st line in our family, so that the number of paths to the point $(n - r, r)$ is the rth entry in the $(n + 1)$st line of the above array. But the $(n + 1)$st line contains the binomial coefficients $C(n, 0)$, $C(n, 1)$, ... and the rth entry is $C(n, r)$.

An alternative proof of Theorem 1 can be obtained by noting that the required number of paths is the same as the number of permutations of n objects with $n - r$ alike of one kind and r alike of a second kind. This number is $n!/(n - r)!r!$ by Theorem 1 of Section 2.2. (See also Problems 10 and 13 of Exercise 2.2.)

In Theorem 2 of Section 2.4 we determined the maximum term in the expansion of $(x + a)^n$ with $ax > 0$. In the following theorem we obtain the maximum value of the binomial coefficients $C(n, r)$ for fixed n.

THEOREM 2 For fixed n, the greatest value of the binomial coefficient $C(n, r)$ occurs for $r = [(n + 1)/2]$.

Proof The theorem may be derived from Theorem 2 of Section 2.4 by setting $x = a = 1$ in the result of that theorem. The proof can be given directly. It is left to the reader to verify that

$$
C(n, r) = \left(\frac{n + 1}{r} - 1 \right) C(n, r - 1). \tag{1}
$$

Thus the sequence

$$C(n, 1), \ C(n, 2), \ \ldots$$

is monotone increasing as long as the multiplying factor in (1) is greater than unity, that is, as long as

$$\frac{n + 1}{r} - 1 > 1 \quad \text{or} \quad \frac{n + 1}{2} > r.$$

The theorem follows. In Problem 14 of Exercise 2.5 the reader is asked to investigate the situation in which $(n + 1)/2$ is an integer.

EXAMPLE 4   The maximum coefficient in $(x + y)^{14}$ is $C(14, 7)$ since $(14 + 1)/2 = 7$.

In the following examples we give some identities involving binomial coefficients. We indicate the derivation of the identities but leave the details for the reader.

EXAMPLE 5   Show that

$$C(n, 0) + C(n, 1) + \cdots + C(n, n) = 2^n. \tag{2}$$

Solution   To obtain this identity, set $x = 1$ in the expansion of $(x + 1)^n$.

EXAMPLE 6   Show that

$$C(n, 1) + C(n, 3) + \cdots = C(n, 0) + C(n, 2) + \cdots . \tag{3}$$

Solution   We are asked to show that the sum of the odd binomial coefficients is equal to the sum of the even binomial coefficients. To derive (3), set $x = -1$ in the expansion of $(x + 1)^n$.

EXAMPLE 7   Show that

$$C(n, 0) + 2C(n, 1) + \cdots + (n + 1)C(n, n) = 2^n + n2^{n-1}. \tag{4}$$

Solution   To derive (4), write the left side of (4) in the form

$$C(n, 0) + C(n, 1) + \cdots + C(n, n)$$
$$+ n[C(n - 1, 0) + C(n - 1, 1) + C(n - 1, 2) + \cdots]$$

by using the identity $rC(n, r) = nC(n - 1, r - 1)$ obtained from Problem 7 of Exercise 2.4, and then apply (2) twice.

A simpler derivation of this identity can be obtained by using the identities of Example 5 above and Problem 16(a) of Exercise 2.5.

EXAMPLE 8   Show that

$$\binom{n}{1}^2 + 2\binom{n}{2}^2 + 3\binom{n}{3}^2 + \cdots + n\binom{n}{n}^2 = \frac{(2n - 1)!}{((n - 1)!)^2}. \tag{5}$$

Solution   To obtain (5), we proceed as follows. Differentiate the expansion of $(1 + x)^n$

$$n(1 + x)^{n-1} = C(n, 1) + 2C(n, 2)x + \cdots + nC(n, n)x^{n-1}$$

so that

$$nx(1 + x)^{n-1} = C(n, 1)x + 2C(n, 2)x^2 + \cdots + nC(n, n)x^n,$$

and then replace x by $1/x$ to obtain

$$\frac{n}{x}\left(1 + \frac{1}{x}\right)^{n-1} = \frac{C(n, 1)}{x} + \frac{2C(n, 2)}{x^2} + \cdots + \frac{nC(n, n)}{x^n}.$$

Now multiply this sum by

$$(1 + x)^n = 1 + C(n, 1)x + \cdots + C(n, n)x^n.$$

The coefficient of x^0 is the sum of the terms on the left of (5) and this is then the coefficient of x^0 in the expansion of

$$\frac{n}{x}(1 + x)^n\left(1 + \frac{1}{x}\right)^{n-1} \qquad \text{or} \qquad \frac{n}{x^n}(1 + x)^{2n-1},$$

that is, the coefficient of x^n in the expansion of $n(1 + x)^{2n-1}$.

Finally, it was shown in Section 1.1 that the number of subsets of a set containing n elements is 2^n, which is precisely the sum of the binomial coefficients as shown in (2) above. This yields an alternative proof of the following theorem.

THEOREM 3   The number of subsets of a set of n objects is 2^n.

Proof   The number of subsets of n objects containing r objects is $\binom{n}{r}$. But each subset contains either 0, 1, 2, $\ldots$, or n objects so that the number of subsets is, from (2),

$$\binom{n}{0} + \binom{n}{1} + \cdots + \binom{n}{n} = 2^n.$$

EXERCISE 2.5

1. Use Pascal's formula to prove the binomial theorem, for n a positive integer, by induction.

2. Derive $C(n + 1, r) = C(n, r) + C(n, r - 1)$, (a) from first principles, (b) using Pascal's formula.

3. Prove that $C(n + 1, r) = C(n, r - 1) + C(n - 1, r) + C(n - 1, r - 1)$.

4. Consider the $C(n, r)$ r-subsets of n distinct objects and select two particular objects P and Q. Then the r-subsets fall into four categories, those containing both P and Q, those with P but not Q, etc. Write $C(n, r)$ as the sum of the numbers of members of these four classes.

5. Solve for n: $C(n + 1, 4) = C(n, 3)$.

6. Prove that $C(n, r) + C(n, r + 1) = C(n + 1, r + 1)$.

7. Show that the sum of the elements of the eighth row of Pascal's triangle is 2^7.

8. Generalize the result of Problem 7 to the nth row.

9. Show that the sum of the elements of the nth row of Pascal's triangle equals one plus the sum of the elements of all previous rows.

10. Draw the 15 possible paths from the origin to the point $(4, 2)$ in Example 3.

11. Find the number of distinct paths from the origin to the point $(8, 5)$ that a particle can take if it can move only one unit in the positive x or y direction at a time.

12. A man wishes to reach an intersection six blocks east and four blocks north of him in the shortest possible time. How many routes can he travel?

13. Find r for which the binomial coefficient $C(12, r)$ is greatest.

14. Show that if n is odd, the greatest value of the binomial coefficient $C(n, r)$ occurs for both $(n + 1)/2$ and $(n - 1)/2$.

15. Complete the derivations of the identities (2)–(5) of this section (Examples 5–8).

16. Prove the following identities:

(a) $\dbinom{n}{1} + 2\dbinom{n}{2} + \cdots + n\dbinom{n}{n} = n2^{n-1}$.

(b) $1 + \dfrac{1}{2}\dbinom{n}{1} + \dfrac{1}{3}\dbinom{n}{2} + \cdots + \dfrac{1}{n+1}\dbinom{n}{n} = \dfrac{2^{n+1} - 1}{n + 1}$.

(c) $\binom{n}{0} + 2\binom{n}{1} + \binom{n}{2} + 2\binom{n}{3} + \binom{n}{4} + 2\binom{n}{5} + \cdots = 3 \cdot 2^{n-1}$.

(d) $\dfrac{1}{2}(1 + x)^n + \dfrac{1}{2}(1 - x)^n = 1 + \binom{n}{2}x^2 + \binom{n}{4}x^4 + \cdots$.

(e) $\dfrac{\binom{n}{1}}{\binom{n}{0}} + \dfrac{2\binom{n}{2}}{\binom{n}{1}} + \dfrac{3\binom{n}{3}}{\binom{n}{2}} + \dfrac{4\binom{n}{4}}{\binom{n}{3}} + \cdots + \dfrac{n\binom{n}{n}}{\binom{n}{n-1}} = \dfrac{1}{2}n(n + 1)$.

(f) $\binom{n}{1} - 2\binom{n}{2} + 3\binom{n}{3} - 4\binom{n}{4} + \cdots + (-1)^{n-1}n\binom{n}{n} = 0$.

(g) $\binom{n}{0} - \dfrac{1}{2}\binom{n}{1} + \dfrac{1}{3}\binom{n}{2} - \dfrac{1}{4}\binom{n}{3} + \cdots + \dfrac{(-1)^n}{n+1}\binom{n}{n} = \dfrac{1}{n+1}$.

(h) $\binom{n}{0} + 3\binom{n}{1} + 5\binom{n}{2} + 7\binom{n}{3} + \cdots = 2^n(n + 1)$.

(i)* $\binom{n}{0}\binom{n}{1} + \binom{n}{1}\binom{n}{2} + \binom{n}{2}\binom{n}{3} + \cdots + \binom{n}{n-1}\binom{n}{n}$

$$= \dfrac{(2n)!}{(n + 1)!\,(n - 1)!}.$$

(j)* $\binom{n}{0}^2 + \binom{n}{1}^2 + \binom{n}{2}^2 + \cdots + \binom{n}{n}^2 = \dfrac{(2n)!}{(n!)^2}$.

(k)* $\binom{n}{0}\binom{n}{r} + \binom{n}{1}\binom{n}{r+1} + \binom{n}{2}\binom{n}{r+2} + \cdots + \binom{n}{n-r}\binom{n}{n}$

$$= \dfrac{(2n)!}{(n - r)!(n + r)!}.$$

(l)* $2\binom{n}{0} + \dfrac{2^2}{2}\binom{n}{1} + \dfrac{2^3}{3}\binom{n}{2} + \dfrac{2^4}{4}\binom{n}{3} + \cdots = \dfrac{3^{n+1} - 1}{n + 1}$.

17.   Prove that in any row of Pascal's triangle the sum of the odd-numbered elements is equal to the sum of the even-numbered elements.

18.   Find the sum

$$1 + 2 + 3 + \cdots + n$$

in the following way. Write Pascal's Formula as

$$C(m - 1, r - 1) = C(m, r) - C(m - 1, r).$$

Choose $r = 2$ and replace m in turn by $n + 1, n, n - 1, \ldots, 4, 3$. Add the resulting equations.

19. Show that

$$1 \cdot 2 + 2 \cdot 3 + 3 \cdot 4 + \cdots + n(n + 1) = \tfrac{1}{3}n(n + 1)(n + 2)$$

by repeating the procedure in (18) but raising all values of m and r by 1; that is, replace r by 3 and m in turn by $m + 2, m + 1, m, \ldots, 5, 4$ in the equation in (18). Add the resulting equations and substitute the values for the combination symbols.

20. In the result of (19), write

$$1 \cdot 2 = 1(1 + 1) = 1^2 + 1$$
$$2 \cdot 3 = 2(2 + 1) = 2^2 + 2$$
$$3 \cdot 4 = 3(3 + 1) = 3^2 + 3$$
$$\vdots$$

and develop the formula for

$$1^2 + 2^2 + \cdots + n^2.$$

21.[c] Extend Pascal's triangle to $n = 100$ and hence obtain the expansion for $(x + 1)^{100}$.

2.6 The Multinomial Theorem

This theorem is an extension of the binomial theorem.

MULTINOMIAL THEOREM   If n is a positive integer, then

$$(x_1 + x_2 + \cdots + x_k)^n = \sum \binom{n}{r_1, r_2, \ldots, r_k} x_1^{r_1} x_2^{r_2} \cdots x_k^{r_k},$$

where the summation is taken over all nonnegative integral values of $r_1, \ldots, r_k$ such that $r_1 + r_2 + \cdots + r_k = n$, and where

$$\binom{n}{r_1, r_2, \ldots, r_k} = \frac{n!}{r_1! r_2! \cdots r_k!}.$$

We do not prove this theorem, but it may be proved in a manner similar to the proof of the binomial theorem, as indicated in Problems 1–3 of Exercise 2.6, or by induction on k, as required in Problem 6.

COROLLARY   The term involving $\alpha^r \beta^s \gamma^t \cdots$ in the expansion of

$$(\alpha + \beta x + \gamma x^2 + \cdots)^p$$

is

$$\frac{p!}{r!\,s!\,t!\,\cdots}\, \alpha^r \beta^s \gamma^t \cdots x^{s+2t+\cdots}$$

where $r + s + t + \cdots = p$.

EXERCISE 2.6

1. Show from first principles that the coefficient of the term $a^3 b^5 c^2 d^3$ in the expansion of $(a + b + c + d)^{13}$ is

$$\frac{13!}{3!\,5!\,2!\,3!}.$$

2. Show from first principles that the coefficient of the term $x^2 y^5 z^3 t u^3$ in the expansion of $(x + y + z + t + u)^{14}$ is

$$\frac{14!}{2!\,5!\,3!\,1!\,3!}.$$

3. Show that the coefficient of the term $x_1^{r_1} x_2^{r_2} \cdots x_k^{r_k}$ with

$$r_1 + r_2 + \cdots + r_k = n$$

in the expansion of $(x_1 + x_2 + \cdots + x_k)^n$ is

$$\frac{n!}{r_1!\,r_2!\,\cdots r_k!}.$$

4. What is the sum of the coefficients in the expansion of $(a + b + c)^6$? of $(a + b + c + d)^{13}$? of $(x + y + z + t + u)^{14}$?

5. What is the sum of all numbers of the form

$$\frac{10!}{x!\,y!\,z!}$$

where x, y, z are nonnegative integers whose sum is 10?

6.* Prove the multinomial theorem by induction on k.

7. Find the coefficient of x^5 in the expansion of $(a + bx + cx^2)^9$.

8. Find the sum $1^3 + 2^3 + \cdots + n^3$.

9. Prove that

$$\sum_{r+s+t=n} \frac{n!}{r!\,s!\,t!} = 3^n.$$

2.7 Stirling's Formula

An examination of a table of factorials indicates that $n!$ increases very rapidly with n. Since computations are usually carried out to some fixed degree of accuracy, for example, to four significant figures, it is often sufficient to use an approximation to $n!$ in a computation. A simple approximation to $n!$ is given by Stirling's formula.

STIRLING'S FORMULA $n! \sim S(n)$ where $S(n) = \sqrt{2\pi}\, n^{n+\frac{1}{2}} e^{-n}$. Note: The expression $f(n) \sim g(n)$ means that

$$\lim_{n \to \infty} \frac{f(n)}{g(n)} = 1.$$

EXAMPLES

n	$n!$	$S(n)$	percentage error
1	1	0.9221	8
2	2	1.919	4
5	120	118.019	2
10	3,628,800	3,598,600	0.8

Stirling's approximation of $n!$ is easy to use on the computer. The above examples show that the approximation is good. But it is important to know how good. A precise estimate is given in the following theorem.

THEOREM $n! = \sqrt{2\pi}\, n^{n+\frac{1}{2}} e^{-n} \cdot e^{r_n}$

where

$$\frac{1}{12n+1} < r_n < \frac{1}{12n}.$$

This theorem was proved by A. Robins [A remark on Stirling's formula, *American Mathematical Monthly*, **62** (1955), 26–29]. Proofs of Stirling's

Formula may be found in this paper, and in a paper by W. Feller [A direct proof of Stirling's formula, *American Mathematical Monthly*, **74** (1967), 1223–1225].

E X E R C I S E 2.7

1. In the theorem of this section, show that

$$r_n = \log(n!/S(n)).$$

2.^{c} Write a computer program to evaluate $n!$.

3.^{c} Write a computer program to evaluate $S(n)$.

4.^{c} Use the computer programs of Problems 2 and 3 and the formula for r_n in Problem 1 to verify the validity of the theorem of this section for $n \leq 20$.

The Inclusion–Exclusion Principle

3.1 A Calculus of Sets

We assume that the reader is familiar with elementary set theory, in particular with set notation: the universal set U in a given situation; the idea of set inclusion, $A \subset B$ (set A is contained in set B); the binary operations of intersection $A \cap B$, or AB, and of union $A \cup B$ of two sets A and B; and the unary operation of complementation (A' is the complement of the set A in U).

In elementary set theory we normally deal with sets of distinct elements. In this section we consider manipulations with sets in which repetitions are allowed. For example, the set $[a, a]$ containing two elements a will be considered as different from the set $\{a\}$ containing the single element a. The symbol [ ] will be used instead of { } to emphasize the distinction.

The symbols $+$ and $-$ will be used to generate sets with repeated elements. The symbol $A + B$ represents the totality of elements in A and B (with repeats counted); $A - B$ is the set of elements in A and not in B, and the symbol will be used only if B is a subset of A. If U is the universal set containing A, then $U - A$ is the **relative complement** or **complement** of the set A in the set U. We use the symbol $'$ to denote relative complement, that is,

$A' = U - A$, $B' = U - B$, etc., and we sometimes use Greek miniscules to refer to the same sets, $\alpha = A'$, $\beta = B'$, etc.

EXAMPLE 1   If $U = \{1, 2, 3, \ldots, 10\}$ $A = \{1, 2, 3\}$, $B = \{2, 3, 4, 5\}$, and $C = \{1, 2, 3, 4, 6\}$, then

$$A + B = [1, 2, 2, 3, 3, 4, 5], \qquad B + C = [1, 2, 2, 3, 3, 4, 4, 5, 6],$$
$$C - A = \{4, 6\}, \qquad \gamma = C' = U - C = \{5, 7, 8, 9, 10\}.$$

In a set-theoretic expression, $A' = U - A$ can be considered to act just like an algebraic factor in the expression. For example, from Figure 3.1, the reader who is familiar with set theory can check that, if A and B are subsets of U, then

$$\alpha\beta = A'B' = U - A - B + AB.$$

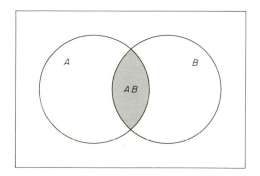

Figure 3.1

The set $A'B'$ is the set of elements common to A' and B', that is, the set of elements that are both not in A and not in B. If we subtract the elements of A and B from U, we have subtracted the elements in AB twice, and so we add these elements back in. Now if we proceed algebraically, we write

$$\begin{aligned}
A'B' &= (U - A)(U - B) \\
&= UU - AU - UB + AB \\
&= U - A - B + AB
\end{aligned}$$

since $UU = U$, $AU = A$, $UB = B$, and get the same result as above.

EXAMPLE 2   If A, B, and C are subsets of U, then

$$\begin{aligned}
A'B'C' &= (U - A)(U - B)(U - C) \\
&= UUU - AUU - UBU - UUC + ABU + AUC + UBC - ABC \\
&= U - A - B - C + AB + AC + BC - ABC,
\end{aligned}$$

a result that can be checked by set-theoretic arguments.

It is true that this kind of symbolic operation is possible for any finite number of sets, however we will not prove this statement.

EXAMPLE 3   If A, B, and C are subsets of U, then

$$
\begin{aligned}
AB(BC')' &= AB(U - BC') \\
&= AB[U - B(U - C)] \\
&= AB[U - B + BC] \\
&= AB - AB + ABC \\
&= ABC.
\end{aligned}
$$

Example 3 illustrates how a complicated set expression can sometimes be simplified algebraically. The result in this example can be checked by set-theoretic arguments.

EXAMPLE 4

(i)   Given a universal set U and one subset A, the sets that can be formed by taking all possible intersections of these two sets are the sets U and A themselves ($UU = U$, $UA = A$, $AA = A$).

(ii)   Given U and two subsets A and B, the sets that are intersections of these are U, A, B, and AB.

(iii)   Given U and three subsets A, B, and C, the following are all possible intersections of these sets:

$$U, \ A, \ B, \ C, \ AB, \ BC, \ CA, \ ABC.$$

In general, given n subsets $A, B, C, \ldots, N$ of a universal set U, we can form 2^n intersections of these sets two or more at a time:

$$U, A, B, \ldots, N, AB, \ldots, ABC, \ldots, ABCD, \ldots, (ABC \ldots N).$$

We call these sets the **positive sets** of U relative to the subsets $A, B, C, \ldots, N$. By a positive set of U with respect to subsets $A, B, \ldots, N$ we mean a set that is the intersection of two or more of the sets $U, A, B, \ldots, N$. In (i) of Example 4 we list the two positive sets of U relative to the subset A; in (ii) we list the four positive sets of U relative to the subsets A and B; in (iii) we list the eight positive sets of U relative to the subsets A, B, and C. Note that if $AB \subset C$, for example, then $ABC = AB$. However, for the purposes of counting we will consider these to be different (formal) positive sets. (This is analogous to considering the equation $x^2 = 0$ to have two roots.)

Relative to the n subsets $A, B, \ldots, N$ of U we refer to the positive sets $A, B, \ldots, N$ as **first-order positive sets**, the sets $AB, BC, \ldots,$ as **second-order positive sets**, etc. In general, an **rth-order positive set** will be the intersection of r of the n given sets $A, B, \ldots, N$.

The reader is asked in Problem 1 of Exercise 3.1 to prove that there are 2^n formal positive sets of U relative to n given subsets. The expression "formal positive sets" will be used to indicate that positive sets which are products of different sets will be considered to be different positive sets even though they may contain the same elements.

EXAMPLE 5   From Figure 3.1 we can check that

$$A \cup B = A + B - AB.$$

This gives an expression for $A \cup B$ in terms of positive sets of U relative to subsets A and B.

EXAMPLE 6   Express (a) $A'B'CD$, (b) $(A'BDE')'$ in terms of positive sets of U relative to subsets A, B, C, D, and E.

(a) $\begin{aligned}[t] A'B'CD &= (U - A)(U - B)CD \\ &= (U - A - B + AB)CD \\ &= CD - ACD - BCD + ABCD. \end{aligned}$

(b) $\begin{aligned}[t] (A'BDE')' &= U - A'BDE' \\ &= U - (U - A)BD(U - E) \\ &= U - BD + ABD + EBD - ABDE. \end{aligned}$

EXERCISE 3.1

1. Prove that there are 2^n (formal) positive sets of U relative to n subsets A, $B, \ldots, N$ (a) by induction; (b) by showing that the recurrence relation $t_{n+1} = 2t_n$ with $t_1 = 2$ holds, and using the method described in Section 1.1.

2. Express

   (a) $A \cup B \cup C$ in terms of positive sets of U relative to subsets A, B, and C;

   (b) $AB'C'D$ in terms of positive sets of U relative to subsets A, B, C, and D.

3. Using the symbolic approach of this section, express the following sets in terms of positive sets of U relative to subsets A, B, C, D, and E;

   (a) A' (b) $A'B'$ (c) $A'B'C'$ (d) $A'BCD'$
   (e) $A'B'CD'$ (f) $ABC'D'E'$ (g) $(AC'DE)'$.

4. Verify the expression for $A'B'C'$ found in Example 2.

5. Prove that

   (a) $A'B' = (A \cup B)'$,   (b) $A'B'C' = (A \cup B \cup C)'$.

6.* Generalize the results in (5) to n sets $A, B, \ldots, N$.

3.2 The Inclusion–Exclusion Principle

The expression

$$A \cup B = A + B - AB$$

found in Example 5 of Section 3.1 tells us that if we wish to have the elements that are in either set A or set B (or both) we count in or *include* the elements in sets A and B but then subtract or *exclude* the set of elements in AB (of course $A + B$ has counted the elements of AB twice).

Again in Section 3.1 we encountered the expression

$$A'B' = U - A - B + AB$$

which tells us that the elements of U which are in neither A nor B can be found by taking the elements of U, subtracting the elements of A and B (this subtracts the elements of AB twice) and adding back in the elements of AB.

Both of these expressions involve adding in certain elements and subtracting certain elements, that is, including certain elements and excluding certain elements. This inclusion–exclusion idea is important in combinatorics and we study it further in this section.

EXAMPLE 1   Let U be a certain set of people. Let A denote the subset of U with the characteristic of having blue eyes and let B denote the subset of males. Let $n(A)$ denote the number of elements in set A. Then, since

$$A'B' = U - A - B + AB,$$

it follows from the definition of addition and subtraction of positive sets that

$$n(A'B') = n(U) - n(A) - n(B) + n(AB).$$

This latter expression, related to the sets of this example, tells us that the number of non-blue-eyed females in the set is the total number of people in the set less the number having blue eyes, less the number of males, plus the number of blue-eyed males (the latter number having been excluded twice in subtracting the number of those people with blue eyes and those who are male).

The result in Example 1 can be generalized as follows. Let U be a set consisting of m elements and consider t attributes or properties or characteristics $P_1, P_2, \ldots, P_t$. Let A be the subset of U possessing the attribute P_1, B the subset of U possessing the attribute $P_2, \ldots, T$ the subset possessing the attribute P_t. Then the number of elements of the set U possessing *none* of the attributes $P_1, P_2, \ldots, P_t$ is given by

$$
\begin{aligned}
n(A'B'C'\cdots T') = m &- [n(A) + n(B) + \cdots n(T)] \\
&+ [n(AB) + n(AC) + \cdots] \\
&- [n(ABC) + \cdots] \\
&+ \cdots + (-1)^t n(ABC\cdots T)
\end{aligned}
\tag{1}
$$

where the bracketed expressions involve, in turn, the sums of the numbers of elements in the first-order, second-order, third-order, $\ldots$, tth-order positive sets of U relative to subsets A, $B, \ldots, T$. The use of formula (1) is sometimes called the application of the **inclusion–exclusion** principle.

We may verify formula (1) as follows. An element with none of the t specified properties is counted once in the number $n(U) = m$ and not in any of the remaining terms in the sum on the right side of (1). An element with exactly one property, say P_1, is counted once in m and once in $n(A)$ and not at all in the remaining terms and so contributes 0 to the right side of (1). An element with exactly h of the properties, say $P_1, P_2, \ldots, P_h$, is counted once on the right side of (1) in each $n(V)$ where V is a positive set of U relative to A, B, $\ldots$, H, and not at all in $n(W)$ for any of the remaining positive sets W of U relative to A, B, $\ldots$, T. Then since the element is counted once in $n(U)$, and since there are $\binom{h}{1}$ first order positive sets, $\binom{h}{2}$ second order positive sets, $\ldots$, and $\binom{h}{h}$ hth order positive sets of U relative to A, B, $\ldots$, H, then the element with exactly the properties $P_1, P_2, \ldots, P_h$ will be counted

$$
1 - \binom{h}{1} + \binom{h}{2} - \binom{h}{3} + - \cdots + (-1)^h \binom{h}{h} = (1-1)^h = 0
$$

times on the right side of (1). We conclude that the expression on the right side of (1) counts only those elements having none of the properties P_1, $P_2, \ldots, P_t$, and this establishes the formula stated in (1).

Since

$$
A'B'C'\cdots T' = (A \cup B \cup C \cup \cdots \cup T)',
$$

as the reader was asked to conjecture and verify in Problem 6 of Exercise 3.1, we have

$$
A'B'C'\cdots T' = U - A \cup B \cup C \cup \cdots \cup T
$$

so that, from (1),

$$n(A \cup B \cup C \cup \cdots \cup T) = n(U) - n(A'B'C' \cdots T')$$
$$= [n(A) + n(B) + \cdots + n(T)]$$
$$- [n(AB) + n(AC) + \cdots]$$
$$+ [n(ABC) + \cdots]$$
$$- \cdots + (-1)^{t-1} n(ABC \cdots T) \qquad (2)$$

where the positive sets on the right are the positive sets on the right of (1) except for U itself. An application of this formula (2) will also be an application of the inclusion–exclusion principle. Here we have an expression for the number of elements possessing at least one of the properties $P_1, P_2, \ldots, P_t$ in terms of the number of elements possessing exactly one, exactly two, $\ldots$, exactly t of the properties.

EXAMPLE 2   Out of 100 people interviewed, 47 smoked, 29 chewed gum, and 18 both smoked and chewed gum. How many were both nonsmokers and nonchewers of gum?

Solution   Let A be the set of smokers and B the set of chewers of gum. Then AB is the set of chewing smokers and $A'B'$ the set of nonsmoking nonchewers.

$$n(A'B') = n(U) - n(A) - n(B) + n(AB)$$
$$= 100 - 47 - 29 + 18 = 42.$$

The required answer is 42.

EXAMPLE 3   In Example 2, how many people either smoked or chewed gum?

Solution   The required number is $n(A \cup B)$ where

$$n(A \cup B) = n(U) - n(A'B')$$
$$= 100 - 42 = 58.$$

EXERCISE 3.2

1.   Let U, A, and B, be defined as in Example 1 of Section 3.1. Compute $n(A'B')$ and $n(A \cup B)$.

2.   Let C, as defined in Example 1 of Section 3.1, be added to the sets of Problem 1. Compute

   (a) $n(B'C')$ (b) $n(A \cup C)$
   (c) $n(A'B'C')$ (d) $n(A \cup B \cup C)$.

3. In a certain factory there are 800 employees; there are 300 men, 552 union members, 424 married employees, 188 male union members, 166 married men, 208 married union members, and 144 married male union members. Find the number of single, nonunion female employees.

4. In a recent survey of 400 women, one-half were found to be married, one half owned fur coats, and one-half had had their appendices removed; in addition, there were 104 married women with fur coats, 116 married women with no appendices, 93 women with fur coats and no appendices, and 53 single, unfurred, appendix-owning women. Find the number of married, furred, de-appendicized women.

3.3 Some Applications of the Inclusion–Exclusion Principle

1. NUMBER THEORY

EXAMPLE 1 How many integers between 1 and 600 inclusive are (a) not divisible by 3? (b) not divisible by either 3 or 5? (c) not divisible by 3, 5 or 7?

Solution (a) Let U be the set of integers between 1 and 600 inclusive so that $m = n(U) = 600$. Let A be the set of integers in U divisible by 3. Since every third integer is divisible by 3, then $n(A) = \frac{1}{3}(600) = 200$. It follows that

$$n(A') = m - n(A) = 600 - 200 = 400,$$

which gives the required number.

(b) Let B be the set of integers in U divisible by 5; then $n(B) = \frac{1}{5}(600) = 120$. The required number is

$$n(A'B) = m - n(A) - n(B) + n(AB)$$

where $n(AB)$ is the number of integers in U divisible by both 3 and 5, that is, by 15. Then $n(AB) = 600/15 = 40$ and

$$n(A'B') = 600 - 200 - 120 + 40 = 320.$$

(c) Let C be the set of integers in U divisible by 7; then $n(C) = [600/7]$ since $n(C)$ is an integer. Then $n(C) = 85$ and the reader should check that $n(BC) = 17$, $n(CA) = 28$, and $n(ABC) = 5$ so that, using (1) of Section 3.2,

$$\begin{aligned}
n(A'B'C') &= m - [n(A) + n(B) + n(C)] \\
&\quad + [n(AB) + n(BC) + n(CA)] \\
&\quad - n(ABC) \\
&= 600 - (200 + 120 + 85) + (40 + 17 + 28) - 5 \\
&= 275.
\end{aligned}$$

2. LINEAR EQUATIONS WITH UNIT COEFFICIENTS

EXAMPLE 2   Find the number of solutions in positive integers of the equation

$$x + y + z = 7$$

with $x \leq 3$, $y \leq 4$, and $z \leq 5$.

Solution   Let U be the set of all solutions in positive integers of the equation; let A be the set of solutions in positive integers with $x > 3$, B the set with $y > 4$, and C the set with $z > 5$. Then AB is the set of solutions in positive integers with $x > 3$ *and* $y > 4$, etc. From the inclusion–exclusion principle, the required number of solutions is

$$n(A'B'C') = n(U) - [n(A) + n(B) + n(C)]$$
$$+ [n(AB) + n(BC) + n(CA)] - n(ABC).$$

In Chapter 4 we will find a formula to determine $n(U)$, $n(A)$, $n(AB)$, etc. in such problems but in this simple example it is easy enough to list all of the required solutions to find

$$n(U) = 15, n(A) = 3, n(B) = 1, n(C) = n(AB) = n(BC) = n(CA) = n(ABC) = 0$$

so that the required number is

$$n(A'B'C') = 15 - (3 + 1) = 11.$$

3. CHROMATIC POLYNOMIALS

Chromatic polynomials were introduced in Section 1.4. The inclusion–exclusion principle can be used to obtain the chromatic polynomial of a map as illustrated in Example 3. However, if the number of regions or faces is large, the number of computations required may be too large to handle even with the aid of a computer.

EXAMPLE 3   Find the chromatic polynomial of the map M_4 of Figure 1.19, Chapter 1, which is repeated in Figure 3.2.

Solution   Let U denote the set of *all* colorings of the map M_4 in λ colors, either proper colorings (no neighboring regions with the same color) or improper. Then $n(U) = \lambda^4$. Let A be the set of colorings of U with regions a and b colored the same. Similarly, let B, C, and D denote the sets of colorings of U in which b and c, c and d, and d and a respectively are colored alike. (Notice that A, B, C, and D correspond to the edges of the

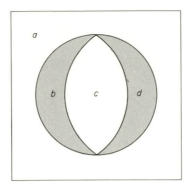

Figure 3.2

corresponding graph. We want to exclude colorings in which adjacent vertices are colored alike.) Then the reader may check that

$$n(A) = n(B) = n(C) = n(D) = \lambda^3$$
$$n(AB) = n(BC) = n(CD) = n(DA) = n(AC) = n(BD) = \lambda^2,$$

and the number of elements in each of the third- and fourth-order positive sets is λ. Then

$$n(A'B'C'D') = n(U) - \sum n(A) + \sum n(AB) - \sum n(ABC) + n(ABCD),$$

where the summations are over all positive sets of the indicated order; thus

$$n(A'B'C'D') = \lambda^4 - 4\lambda^3 + 6\lambda^2 - 4\lambda + \lambda$$
$$= \lambda^4 - 4\lambda^3 + 6\lambda^2 - 3\lambda,$$

as we found in Section 1.4.

EXERCISE 3.3

1. A year is a leap year if it is divisible by four but not by 100, or if it is divisible by 400 (for example, 1900 was not a leap year but 2000 will be). Find the number of leap years from 1884 to 4004 inclusive.

2. How many integers from 1 to 49,000 inclusive are not divisible by 3, 5, or 7?

3. How many integers from 1 to 10,000 inclusive are divisible by none of 5, 7, or 11?

4. How many integers from 1 to 1,000,000 inclusive are neither perfect squares, perfect cubes, nor perfect fourth powers?

5. How many solutions in positive integers less than five are there for the equation $x + y + z = 7$?

6. Find the chromatic polynomial of a tetrahedron by the method described in Example 3 of this section.

7. How many integers from 1 to 100 inclusive have no repeated prime factors?

8.^{c} Write a computer subroutine to apply the inclusion–exclusion principle for the t sets A, B, ..., T. Test the program on the examples and problems of Section 3.2.

9.^{c} Write a computer program to find the chromatic polynomial of a map using the subroutine of Problem 8. Test the program on Example 3 of this section.

10.^{c} Find the chromatic polynomial of a cube.

3.4 Derangements

A **derangement** of a set of objects is a permutation of the set in which none of the objects is mapped onto itself. In this section we will apply the inclusion–exclusion principle to determine the number of derangements of a given set of objects.

EXAMPLE 1

(i) $\begin{pmatrix} 1 & 2 & 3 \\ 3 & 1 & 2 \end{pmatrix}$ is a derangement, $\begin{pmatrix} 1 & 2 & 3 \\ 3 & 2 & 1 \end{pmatrix}$ is not.

(ii) $\begin{pmatrix} x_1 & x_2 & x_3 & x_4 \\ x_2 & x_1 & x_4 & x_3 \end{pmatrix}$ is a derangement, $\begin{pmatrix} x_1 & x_2 & x_3 & x_4 \\ x_4 & x_2 & x_1 & x_3 \end{pmatrix}$ is not.

(iii) $\begin{pmatrix} A & B & C & D & E \\ E & D & A & C & B \end{pmatrix}$ is a derangement, $\begin{pmatrix} A & B & C & D & E \\ B & A & E & D & C \end{pmatrix}$ is not.

The following theorem gives us the number of derangements of a set of objects. The reader is asked to prove this theorem in Problem 9 of Exercise 3.4. The method of proof is shown in Example 3.

THEOREM   The number, $D(n)$, of derangements of a set of n objects is

$$D(n) = n!\left[1 - \frac{1}{1!} + \frac{1}{2!} - \frac{1}{3!} + - \cdots + (-1)^n \frac{1}{n!}\right]. \qquad (3)$$

EXAMPLE 2 $D(1) = 0$ since $\begin{pmatrix} a \\ a \end{pmatrix}$ is the only permutation of one object.

$D(2) = 1$ since $\begin{pmatrix} a & b \\ b & a \end{pmatrix}$ is the only derangement of two objects.

$D(3) = 2$ since $\begin{pmatrix} a & b & c \\ c & a & b \end{pmatrix}$ and $\begin{pmatrix} a & b & c \\ b & c & a \end{pmatrix}$ are the only derangements of three objects.

From the formula (3),

$$D(1) = 1! \left[1 - \frac{1}{1!} \right] = 0$$

$$D(2) = 2! \left[1 - \frac{1}{1!} + \frac{1}{2!} \right] = 1$$

$$D(3) = 3! \left[1 - \frac{1}{1!} + \frac{1}{2!} - \frac{1}{3!} \right] = 2.$$

EXAMPLE 3   Find $D(6)$, the number of derangements of the set

$$S = \{1, 2, 3, 4, 5, 6\}.$$

Solution   Let U be the set of all permutations of S. Then $n(U) = 6!$.
Let A be the subset of permutations of U with the property that 1 is mapped
onto itself, B the subset with 2 mapped onto itself, ..., F the subset with 6
mapped onto itself. Then $D(6) = n(A'B'C'D'E'F')$, the number of permutations having none of these properties.
   Now

$$n(A) = n(B) = \cdots = n(F) = 5!$$

The permutations in A are equivalent to the permutations of $\{2, 3, 4, 5, 6\}$,
and these are 5! in number, etc. In the same way

$$n(AB) = n(AC) = \cdots = 4!$$

The permutations in AB are equivalent to the permutations of $\{3, 4, 5, 6\}$
since both 1 and 2 are mapped onto themselves, and these are 4! in number,
etc. Similarly, the number of elements in each third-order positive set will be
3!, in each fourth-order positive set will be 2!, etc. Since there are $C(6, 1)$
first-order positive sets, $C(6, 2)$ second-order positive sets, and so on, using
the inclusion–exclusion principle we may write,

$$D(6) = 6! - C(6, 1)5! + C(6, 2)4! - C(6, 3)3!$$
$$+ C(6, 4)2! - C(6, 5)1! + C(6, 6)0!$$

$$= 6! \left[1 - \frac{C(6, 1)}{6} + \frac{C(6, 2)}{6 \cdot 5} - \frac{C(6, 3)}{6 \cdot 5 \cdot 4} \right.$$

$$\left. + \frac{C(6, 4)}{6 \cdot 5 \cdot 4 \cdot 3} - \frac{C(6, 5)}{6 \cdot 5 \cdot 4 \cdot 3 \cdot 2} + \frac{C(6, 6)}{6!} \right]$$

$$= 6! \left[1 - \frac{1}{1!} + \frac{1}{2!} - \frac{1}{3!} + \frac{1}{4!} - \frac{1}{5!} + \frac{1}{6!} \right]$$

as predicted in formula (3).

EXERCISE 3.4

1. Evaluate $D(4)$ and list the derangements of $\{1, 2, 3, 4\}$.

2. Evaluate $D(5)$ and $D(7)$.

3. Find the number of derangements of $\{x_1, x_2, \ldots, x_{12}\}$ in which the first six elements $x_1, \ldots, x_6$ are mapped onto

    (a) $x_1, x_2, \ldots, x_6$ in some order.
    (b) $x_7, x_8, \ldots, x_{12}$ in some order.

4. Find the number of permutations of $\{A, B, C, D, E, F\}$ in which A and C are not mapped onto themselves.

5. How many permutations of $\{x_1, x_2, \ldots, x_9\}$ have exactly four elements mapped onto themselves?

6. Prove that
$$D(n) - nD(n-1) = (-1)^n \qquad \text{for} \quad n \geq 2.$$

7. Eight letters are taken from their envelopes, read, and replaced in the envelopes at random. In how many ways can this be done so that (a) none, (b) at least one, (c) at least two of the letters will be in the correct envelope?

8. The labels from seven different cans of soup were unfortunately removed and the new clerk replaced them at random. In how many ways can this be done so that (a) none, (b) at least one, (c) all of the cans will be labeled correctly?

9. Prove the theorem of this section.

10. Deduce the theorem of this section from Problem 6.

11.* In how many ways can eight rooks be placed on a chessboard so that none is on the white diagonal and no two are attacking each other?

12. What is the limit of $D(n)/n!$ as n approaches infinity?

Linear Equations with Unit Coefficients

It is possible to identify the solution of many combinatorial problems with the number of solutions in integers of a linear equation with unit coefficients, that is, an equation of the form

$$x_1 + x_2 + \cdots + x_r = m$$

where m and r are integers. We study such solutions in this chapter. An application to the problem of finding combinations when repetitions are allowed is given.

4.1 Solutions Bounded Below

First consider *positive* solutions of linear equations with unit coefficients, that is, solutions $(x_1, \ldots, x_r)$ with x_i integral and $x_i > 0$ for $i = 1, \ldots, r$.

THEOREM 1   The number of solutions in positive integers of the equation

$$x_1 + x_2 + \cdots + x_r = m$$

is $C(m - 1, r - 1)$.

Proof Set m objects in a line and place markers between $r - 1$ pairs of adjacent objects. This can be done in $C(m - 1, r - 1)$ ways and each way divides the given set of m objects into r nonempty subsets. Let the number of objects in the r subsets be $k_1, k_2, \ldots, k_r$ respectively $(k_i > 0)$. Then

$$x_i = k_i, \qquad i = 1, 2, \ldots, r$$

yields a solution $(x_1, x_2, \ldots, x_r)$ of the equation and all solutions can be obtained in this way. This proves the theorem.

This result is approached in a slightly different way in the following example.

EXAMPLE 1   Find the number of solutions in positive integers of the equation

$$x + y + z = 7.$$

Solution Consider the following line of seven k's separated by t's

$$k \, t \, k \, t \, k \, t \, k \, t \, k \, t \, k \, t \, k \, .$$

If we select any two t's and omit the remainder, as in the array

$$k \, k \, t \, k \, t \, k \, k \, k \, k,$$

we divide the seven k's into three subsets. The numbers of k's in these subsets give a solution in positive integers of the given equation, in this case $(2, 1, 4)$; and, conversely, each solution in positive integers of the equation corresponds to a division of the set of k's into three nonempty subsets. Hence any selection of two of the t's will give a solution so that the number of solutions is $C(6, 2)$ or 15.

The number of nonnegative solutions of a linear equation with unit coefficients can be deduced from Theorem 1.

THEOREM 2   The number of nonnegative integral solutions of the equation

$$x_1 + x_2 + \cdots + x_r = m$$

is $C(m + r - 1, r - 1)$ or $C(m + r - 1, m)$.

Proof The transformation $x_i = y_i - 1$, $i = 1, 2, \ldots, r$, yields an equation

$$y_1 + y_2 + \cdots + y_r = m + r.$$

There is a one-to-one correspondence between the solutions in positive integers of this equation and the solutions in nonnegative integers of the given equation. By Theorem 1, the number of solutions is $C(m + r - 1, r - 1) = C(m + r - 1, m)$.

This result can also be obtained by an argument analogous to that used in Theorem 1. This is demonstrated in Example 2.

EXAMPLE 2   Find the number of solutions in nonnegative integers of the equation

$$x + y + z = 7.$$

Solution   Consider a line of seven k's and two t's such as

$$k\,k\,t\,k\,k\,k\,t\,k\,k.$$

The markers t divide the set of k's into three subsets and the number of k's in the three subsets yields a solution to the equation, namely $(2, 3, 2)$. Similarly the array

$$t\,k\,t\,k\,k\,k\,k\,k$$

yields the solution $(0, 1, 6)$. The number of solutions of $x + y + z = 7$ in nonnegative integers is therefore the number of ways of writing seven k's and two t's in a row. This number is $C(9, 2)$ or 36.

EXAMPLE 3   Find the number of integral solutions of the equation

$$x + y + z = 24$$

in which each integer is greater than 1.

Solution   Let (x, y, z) be a solution and let

$$X = x - 1, \qquad Y = y - 1, \qquad Z = z - 1.$$

Then

$$X + Y + Z = 21$$

is equivalent to the given equation; that is, the required number of solutions of the original equation in integers greater than 1 is the number of solutions in positive integers of the equation $X + Y + Z = 21$. This number is $C(20, 1)$ or 190, by Theorem 1.

The general situation in which each variable x_i is bounded below by an integer, positive or negative, can be handled by a transformation of coordinates.

THEOREM 3   The number of integral solutions of the equation

$$x_1 + x_2 + \cdots x_r = m$$

with $x_i > a_i$, $i = 1, 2, \ldots, r$, is $C(m - a_1 - a_2 - \cdots - a_r - 1, r - 1)$.

Proof   Set $x_i = y_i + a_i$, $i = 1, 2, \ldots, r$, to obtain the equation

$$y_1 + y_2 + \cdots + y_r = m - a_1 - a_2 - \cdots - a_r.$$

There is a one-to-one correspondence between the solutions in positive integers of this equation and the solutions of the original equation subject to the prescribed conditions. But the number of solutions in positive integers of the equation in the y_i is $C(m - a_1 - a_2 - \cdots - a_r - 1, r - 1)$ from Theorem 1.

Of course, if $m < \sum a_i$ in Theorem 3, there are no solutions. Theorem 1 is a special case of Theorem 3 with $a_i = 0$ for all i, and Theorem 2 is a special case of Theorem 3 with $a_i = -1$ for all i.

EXAMPLE 4   Find the number of integral solutions of the equation $x + y + z = 24$ in which $x > 1$, $y > 2$, and $z > 3$.

Solution   Let (x, y, z) be a solution of $x + y + z = 24$ and let $x = X + 1$, $y = Y + 2$, $z = Z + 3$. We obtain the equation $X + Y + Z = 18$ and the number of solutions of the original equation subject to the given conditions is the number of solutions of this equation in positive integers, that is, $C(17, 2)$ by Theorem 1.

EXERCISE 4.1

1.   Find the number of solutions of the equation

$$x + y + z + t + w = 36$$

in (a) positive integers, (b) nonnegative integers.

2.   Show that the equations

$$x_1 + \cdots + x_7 = 13 \qquad \text{and} \qquad x_1 + \cdots + x_{14} = 6$$

have the same number of solutions in nonnegative integers.

3.   Show that the equations

$$x_1 + x_2 + x_3 = 12 \qquad \text{and} \qquad x_1 + \cdots + x_{10} = 12$$

have the same number of solutions in positive integers.

4. How many integers between 100 and 1,000,000 have sum of digits (a) equal to five (b) less than five?

5. How many different collections of three coins can be formed if the coins can be either pennies, nickels, dimes, quarters, or half dollars?

6. How many different collections of five coins can be formed in Problem 5?

7. In how many ways can 20 identical balls be placed into six boxes?

8. What is the number in Problem 7 if no box is to remain empty?

9.* If, in Problem 7, the balls are placed in the boxes at random, why is the probability that no box remains empty not equal to the ratio of the answers to Problems 7 and 8?

10. Ten dice are rolled on a table repeatedly. How many different patterns of results can be obtained?

11.* Find the number of different partial derivatives of order r of an analytic function of n variables.

12. Prove Theorem 2 using the method of Example 2.

13. Find the number of solutions of the equation

$$x + y + z + w = 40$$

in integers greater than five.

14. Find the number of solutions of

$$x_1 + x_2 + \cdots + x_6 = 72$$

in positive integers such that (a) $x_1 > 10$, (b) $x_1 > 10$, and $x_5 > 5$.

15. Find the number of solutions in integers greater than -5 of the equation $x + y + z = 1$.

16. What is the number of solutions of

$$x_1 + x_2 + x_3 + x_4 = t$$

in nonnegative integers satisfying

$$x_1 \geq k_1 \quad \text{and} \quad x_2 \geq k_2 ?$$

4.2 Solutions Bounded Above and Below

In Example 2 of Section 3.3 the inclusion–exclusion principle was used to obtain the number of solutions in positive integers of the equation

$$x + y + z = 7 \quad \text{with} \quad x \leq 3, \quad y \leq 4, \quad \text{and} \quad z \leq 5.$$

This number was expressed in terms of the subsets A, B, and C of solutions of the equation in which $x > 3$, $y > 5$, and $z > 6$ respectively, by means of the formula

$$n(A'B'C') = n(U) - [n(a) + n(B) + n(C)] + [n(AB) + n(BC) + n(CA)] - n(ABC)$$

where U denotes the set of all solutions in positive integers of the given equation.

Using the ideas discussed in Section 4.1, we are now in a position to work examples similar to this without having to list all possible solutions to find $n(A)$, $n(AB)$, etc. as we did in Section 3.3.

EXAMPLE 1   Find the number of solutions in positive integers of the equation $x + y + z = 15$ with $x \leq 5$, $y \leq 6$, and $z \leq 8$.

Solution   Let U be the set of solutions in positive integers of the given equation; then $n(U) = C(14, 2) = 91$ from Theorem 1 of Section 4.1. Let A be the set of solutions in positive integers with $x > 5$, B the set with $y > 6$, and C the set with $z > 8$. Then, by Theorem 3 of Section 4.1,

$$n(A) = C(9, 2) = 36, \qquad n(B) = C(8, 2) = 28,$$
$$n(C) = C(6, 2) = 15, \qquad n(AB) = C(3, 2) = 3,$$
$$n(AC) = n(BC) = n(ABC) = 0,$$

so that

$$n(A'B'C') = 91 - (36 + 28 + 15) + 3 = 15,$$

where $A'B'C'$ is the set of solutions in positive integers with $x \leq 5$, $y \leq 6$, and $z \leq 8$.

The result in Example 1 is formulated in general in the following Theorem.

THEOREM 1   Let A_i denote the set of solutions in positive integers of the equation

$$x_1 + x_2 + \cdots + x_r = m$$

such that

$$x_i > b_i, \qquad i = 1, 2, \ldots, r.$$

Then the number, S, of solutions of the equation in positive integers x_i such that $x_i \le b_i$, $i = 1, 2, \ldots, n$ is given by

$$S = n(U) - \sum n(A_1) + \sum n(A_1 A_2) - \cdots + (-1)^r n(A_1 A_2 \cdots A_r)$$

where U is the set of solutions in positive integers of the given equation and the summations are taken over all positive sets of the indicated orders, the positive sets being those of U relative to $A_1, A_2, \ldots, A_r$. Here

$$n(U) = C(m - 1, r - 1).$$

The proof of this theorem follows from an application of the inclusion–exclusion principle as demonstrated in Example 1.

If all the bounds b_i, $i = 1, 2, \ldots, r$, are equal, the number of solutions can be obtained explicitly. We say that the properties $P_1, P_2, P_3, \ldots$, are **symmetric** if the numbers of elements in the jth-order positive sets relative to the sets $A, B, C, \ldots$, having these properties, are equal for $j = 1, 2, 3, \ldots$, that is, if

$$n(A) = n(B) = n(C) = \cdots = n_1,$$
$$n(AB) = n(AC) = n(BC) = \cdots = n_2,$$
$$n(ABC) = n(ABD) = \cdots = n_3, \text{ etc.}$$

This leads to the following lemma.

LEMMA 1   Let $A, B, C, \ldots$, be subsets of U possessing the symmetric properties $P_1, P_2, P_3, \ldots$ respectively. Then the number of elements in the set U having none of the r symmetric properties, $P_1, P_2, P_3, \ldots$, is

$$n(A'B'C' \cdots) = n(U) - C(r, 1)n(A) + C(r, 2)n(AB) - C(r, 3)n(ABC) + \cdots.$$

If we apply Lemma 1 to Theorem 1, we obtain the following result.

THEOREM 2   The number of solutions of

$$x_1 + x_2 + \cdots + x_r = m$$

in positive integers not exceeding a fixed positive integer c is

$$C(m - 1, r - 1) - C(r, 1)C(m - c - 1, r - 1) + C(r, 2)C(m - 2c - 1, r - 1)$$
$$- \cdots + (-1)^r C(r, r)C(m - rc - 1, r - 1).$$

EXAMPLE 2   Find the number of solutions in integers between one and nine inclusive of the equation $x + y + z = 24$.

Solution   The total number of solutions in positive integers is $C(23, 2)$ or
253. Let A be the set of solutions with $x > 9$, B the set with $y > 9$, and C the
set with $z > 9$. These properties are symmetric so we need only compute

$$n(A) = C(14, 2) = 91; \quad n(AB) = C(5, 2) = 10; \quad n(ABC) = C(-4, 2) = 0.$$

The required number of solutions is

$$n(A'B'C') = 253 - C(3, 1)n(A) + C(3, 2)n(AB) - C(3, 3)n(ABC)$$
$$= 253 - 3(91) + 3(10) - 1(0) = 10.$$

Theorems 1 and 2 can be modified to include the case of nonnegative
solutions bounded above. The reader is asked to do this in Problem 9 of
Exercise 4.2
    The number of solutions bounded above and below can also be obtained
by a transformation of coordinates, as in the following theorem.

THEOREM 3   The number of integral solutions of

$$x_1 + x_2 + \cdots + x_r = m$$

such that

$$a_i < x_i \le b_i, \quad i = 1, 2, \ldots, r,$$

with a_i and b_i integers, is the same as the number of solutions of

$$y_1 + y_2 + \cdots + y_r = m - a_1 - a_2 - \cdots - a_r$$

in positive integers y_i bounded above by c_i where

$$c_i = b_i - a_i, \quad i = 1, 2, \ldots, r.$$

This theorem follows by setting

$$y_i = x_i - a_i, \quad i = 1, 2, \ldots, r$$

and applying Theorem 2.

EXERCISE 4.2

1. Find the number of solutions in integers from one to seven inclusive of
the equation

$$x + y + z + t = 16.$$

2. Find the number of solutions in positive integers of the equation

$$x + y + z + t = 23$$

with

$$x \le 9, \qquad y \le 8, \qquad z \le 7, \qquad t \le 6.$$

3. Find the number of solutions of the equation

$$x + y + z + t = 18$$

with

$$1 \le x \le 5, \qquad 1 \le y \le 6, \qquad 2 \le z \le 7, \qquad 4 \le t \le 10.$$

4. In how many ways can a collection of ten coins be made up from a set of eight pennies, seven nickels, four dimes, and three quarters? (Assume that coins of any one denomination are identical.)

5. In Problem 4, how many of the collections contain no dimes?

6. Show that the equations

$$x + y + z + t = 26 \qquad \text{and} \qquad x + y + z + t = 14$$

have the same number of solutions in integers between one and nine inclusive.

7. If r, m, c_1, c_2, and c_3 are positive integers, find a formula for the number of solutions in positive integers of

$$x_1 + x_2 + \cdots + x_r = m$$

with

$$x_1 \le c_1, \qquad x_2 \le c_2, \qquad x_3 \le c_3.$$

8. Find the number of seven-digit positive integers in which the sum of the digits is 15.

9. State and prove theorems analogous to Theorems 1 and 2 for non-negative integral solutions bounded above.

10. Find the solutions of

$$x + y + z = 24$$

in integers such that

$$1 \le x \le 5, \qquad 12 \le y \le 18, \qquad -1 \le z \le 12.$$

4.3 Combinations with Repetitions

In our discussion of combinations of objects in Chapter 2, we normally allowed no repetitions. The following theorem considers combinations with repetitions and relates the result to our work on linear equations with unit coefficients.

· THEOREM   The number of combinations, m at a time, of r distinct objects each available in unlimited supply, is $C(m + r - 1, m)$ or $C(m + r - 1, r - 1)$.

Proof   Let x_i denote the number of objects of type i selected. $i = 1, 2, \ldots, r$. Then the required number is the number of solutions of

$$x_1 + x_2 + \cdots + x_r = m$$

in nonnegative integers. By Theorem 2 of Section 4.1, this number is $C(m + r - 1, r - 1) = C(m + r - 1, m)$, as required.

EXAMPLE 1   Given an unlimited number of red, green, blue, and black balls, how many different combinations of three balls can be selected?

Solution   Let x_1, x_2, x_3, x_4 be the numbers of red, green, blue, and black balls respectively selected. Then

$$x_1 + x_2 + x_3 + x_4 = 3$$

and we require the number of solutions in nonnegative integers for this equation. This number is $C(6, 3)$ or 20.

If, in a given selection problem, the supply is limited for each type of object selected, the problem is equivalent to finding bounded solutions to linear equations. The following example illustrates the general method.

EXAMPLE 2   A bag contains nine white balls, four red balls, and seven black balls. In how many ways can a selection of five balls be made?

Solution   In any given selection, let x be the number of white balls, y the number of red, and z the number of black balls. Then we require the number of solutions in nonnegative integers of the equation $x + y + z = 5$ with $x \leq 9$, $y \leq 4$, and $z \leq 7$.

The total number of solutions in nonnegative integers is $C(7, 2)$ by Theorem 2 of Section 4.1. Let A be the set of solutions with $x \geq 10$, B the set with $y \geq 5$, and C the set with $z \geq 8$. Then the required number is

$$n(A'B'C') = 21 - [n(A) + n(B) + n(C)] + [n(AB) + n(BC) + n(CA) - n(ABC)$$
$$= 21 - 1 = 20$$

since $n(B) = C(2, 2) = 1$ and all other terms in the formula are zero.

EXERCISE 4.3

1. Find the number of combinations of ten objects taken three at a time, each available in unlimited supply.

2. Suppose in Problem 1 only two duplicates of each type of object are available. How many combinations are there in this case of the ten objects taken three at a time?

Recurrence Relations

5.1 Recurrence Relations

Recurrence relations were introduced in Chapter 1 and were also encountered in Chapter 2. The tower of Hanoi puzzle in Section 1.1, for example, involved the recurrence relation

$$S_{n+1} = 2S_n + 1. \tag{1}$$

In this situation the recurrence relation related the minimum number of moves required to transfer a tower of $n + 1$ rings in the puzzle to the minimum number of moves required to transfer a tower of n rings. In the following we recall some other examples of recurrence relations that we have encountered.

The recurrence relation

$$f(n + 1) = f(n) + n + 1 \tag{2}$$

of Section 1.2, related the number of regions into which the plane was divided by $n + 1$ lines in general position to the number into which the plane was divided by n lines in general position.

In Section 2.2,

$$P(n, r) = (n - r + 1)P(n, r - 1) \qquad (3)$$

related the number of r-arrangements of n distinct objects to the number of $(r - 1)$-arrangements of n distinct objects.

The expression

$$C(n, r) = C(n - 1, r) + C(n - 1, r - 1), \qquad (4)$$

known as Pascal's formula, was introduced in Section 2.5 to relate the binomial coefficient $C(n, r)$ to the binomial coefficients $C(n - 1, r)$ and $C(n - 1, r - 1)$.

Another recurrence relation which will be considered several times in future chapters is

$$F(n) = F(n - 1) + F(n - 2). \qquad (5)$$

This is a very famous recurrence relation; it defines a family of sequences (depending on initial conditions) known as Fibonacci sequences. These have been, and continue to be, very widely studied. This recurrence relation is discussed in Section 5.4, and a special chapter, Chapter 15, is devoted to Fibonacci sequences.

In recurrence relations (1), (2), and (5) above, one parameter is involved; in (3) and (4) two parameters are involved.

EXERCISE 5.1

1. Determine a recurrence relation for each of the functions defined below.

(a) $f(n)$ is the number of regions into which the plane is divided by n circles each pair of which intersect in exactly two points and no three meet in a single point.

(b) $f(n)$ denotes the number of regions into which the surface of a sphere is divided by n great circles, no three concurrent. (A great circle of a sphere is the intersection of the sphere and a plane passing through the center of the sphere.)

(c) In a singles tennis tournament, $2n$ players are paired off in n matches; $f(n)$ denotes the number of different ways in which this pairing can be done.

(d)* $f(n)$ denotes the number of solid regions into which three-dimensional space is divided by n planes in general position (every three planes meet in exactly one point but no four planes do).

(e)* $f(n, r)$ is the number of ways of selecting r integers from the ordered set $\{1, 2, \ldots, n\}$ so that consecutive integers are not selected.

2.   Obtain a recurrence relation for the coefficients, a_n in the power series expansions of the following functions.

(a) $f(x) = \dfrac{1}{1 + 3x} = a_0 + a_1x + a_2 x^2 + \cdots$

(b) $f(x) = \dfrac{1}{1 - x - x^2} = a_0 + a_1x + a_2 x^2 + \cdots$

(c) $f(x) = e^x = a_0 + a_1x + a_2 x^2 + \cdots$.

3.   If a_i are positive integers, the symbol

$$x = a_0 + \cfrac{1}{a_1 + \cfrac{1}{a_2 + \cfrac{1}{a_3 + \cfrac{1}{a_4 + \cdots}}}}$$

is called a **simple continued fraction.** It produces a sequence $\{x_n\}$ of partial sums as indicated:

$$x_0 = a_0$$

$$x_1 = a_0 + \cfrac{1}{a_1}$$

$$x_2 = a_0 + \cfrac{1}{a_1 + \cfrac{1}{a_2}}$$

$$x_3 = a_0 + \cfrac{1}{a_1 + \cfrac{1}{a_2 + \cfrac{1}{a_3}}}$$

$$\vdots$$

Obtain a recurrence relation for x_n by proving the following.

(a)   If $x_n = p_n/q_n$ where p_n and q_n are positive integers, then

$$p_n = a_n p_{n-1} + p_{n-2}, \qquad q_n = a_n q_{n-1} + q_{n-2}$$

where $n \geq 2$.

(b) $x_n = x_{n-1} + \dfrac{(-1)^{n-1}}{q_{n-1}q_n}$.

(c) $x_n = x_{n-2} + \dfrac{(-1)^n a_n}{q_n q_{n-2}}$.

5.2 Solution by Iteration

If we stipulate the initial condition $s_1 = 1$ for the recurrence relation $s_{n+1} = 2s_n + 1$, we may calculate successively, or iteratively, $s_2 = 2s_1 + 1 = 3$, $s_3 = 2s_2 + 1 = 7$, $s_4 = 2s_3 + 1 = 15$, etc. The terms of the sequence defined by this relation (with the initial condition) can be obtained sequentially up to any value of n. If a different initial condition is given, a different sequence is defined. For example, if $s_1 = 0$, then $s_2 = 1$, $s_3 = 3$, $s_4 = 7$, etc.

This sequential or iterative method for finding a given term in a sequence can easily be used with a computer to evaluate s_n for very large values of n. This method, however, does not produce a general formula for the nth term. In Section 1.2 we indicated a procedure for determining a possible formula for the nth term of a given sequence. Once such a formula is available, we can try to prove or disprove it by induction. If it is correct, if can be used to compute s_n directly without first evaluating $s_1, s_2, \ldots, s_{n-1}$. The sequence dealt with in this way in Section 1.2 was the sequence defined by

$$f(k + 1) = f(k) + 1, \quad \text{with} \quad f(1) = 2.$$

In the following examples we demonstrate this procedure for three other sequences.

EXAMPLE 1   Conjecture a general term for the sequence defined by $s_{k+1} = s_k + k + 1$ with $s_1 = 1$.

Solution   We calculate the first few terms of the sequence by iteration and look for a pattern in the terms. Such a pattern can be found, in this case, by dividing the nth term by n. (In other situations we might relate the nth term to the nth term of a known sequence, as in Example 2 below.) Consider the results shown in Table 5.1. From these results we conjecture that

$$\frac{s_n}{n} = \frac{n + 1}{2}$$

so that $s_n = \frac{1}{2}n(n + 1)$. This conjecture can be verified by induction.

TABLE 5.1

n	s_n	s_n/n
1	1	$1 = \frac{2}{2}$
2	3	$\frac{3}{2}$
3	6	$2 = \frac{4}{2}$
4	10	$\frac{5}{2}$
5	15	$3 = \frac{6}{2}$
6	21	$\frac{7}{2}$
7	28	$4 = \frac{8}{2}$

EXAMPLE 2   Conjecture a general term for the sequence defined by $t_{k+1} = t_k + \frac{1}{2}(k + 1)(k + 2)$ with $t_1 = 1$.

Solution   Again we calculate the first few terms of the sequence. This time we can discern no obvious pattern to the terms t_n/n so we try something else. If we calculate t_n/s_n with s_n from Example 1 (see Table 5.2) we notice a pattern. We conjecture that

$$\frac{t_n}{s_n} = \frac{n + 2}{3}$$

so that $t_n = \frac{1}{6}n(n + 1)(n + 2)$ using the expression for s_n from Example 1. This conjecture can be proved by induction.

TABLE 5.2

n	t_n	t_n/n	s_n	t_n/s_n
1	1	1	1	$1 = \frac{3}{3}$
2	4	2	3	$\frac{4}{3}$
3	10	$\frac{10}{3}$	6	$\frac{5}{3}$
4	20	5	10	$2 = \frac{6}{3}$
5	35	7	15	$\frac{7}{3}$
6	56	$\frac{28}{3}$	21	$\frac{8}{3}$
7	84	14	28	$3 = \frac{9}{3}$

EXAMPLE 3   Conjecture a general term for the sequence defined by $f(n + 1) = f(n) + (n + 1)^2, f(1) = 1$.

Solution   We again compare $f(n)$ with s_n from Example 1. The work is shown in Table 5.3. We conjecture that

$$\frac{f(n)}{s_n} = \frac{2n + 1}{3}$$

so that

$$f(n) = \frac{1}{6}n(n + 1)(2n + 1)$$

using s_n from Example 1. The reader may verify this conjecture by induction.

TABLE 5.3

n	$f(n)$	s_n	$f(n)/s_n$
1	1	1	$1 = \frac{3}{3}$
2	5	3	$\frac{5}{3}$
3	14	6	$\frac{7}{3}$
4	30	10	$3 = \frac{9}{3}$
5	55	15	$\frac{11}{3}$
6	91	21	$\frac{13}{3}$
7	140	28	$5 = \frac{15}{3}$

EXERCISE 5.2

Determine a number of particular solutions to the following recurrence relations in (1)–(10) by iteration. Conjecture a formula for the general term
defined by each recurrence relation and prove the formula by induction.

1. $s_{n+1} = 2s_n + 1$, $s_0 = 2$.

2. $f(n + 1) = f(n) + n + 1$, $f(1) = 0$.

3. $P(n, r) = (n - r + 1)P(n, r - 1)$, $P(n, 0) = 1$ for all n.

4. $t_{n+1} = t_n + (n + 1)^3$, $t_1 = 1$.

5. $t_{n+1} = t_n + (3n + 2)$, $t_1 = 2$.

6. $t_{n+1} = t_n + \dfrac{1}{(4n + 1)(4n + 5)}$, $t_1 = \frac{1}{5}$.

7. $t_{n+1} = t_n + 2^n$, $t_1 = 1$.

8. $t_{n+1} = t_n + (2n + 1)$, $t_1 = 1$.

9. $t_{n+1} = t_n + (4 - n)$, $t_1 = 4$.

10.* $t_{n+1} = t_n + (n + 1)(n + 2)(n + 3)$, $t_1 = 6$.

11.* Let $T_{n+1}^2 = 1 + \displaystyle\prod_{k=1}^{n} T_k$ with $T_1 = 2$.

Prove that

(a) $T_{n+1} = T_n^2 - T_n + 1$;

(b) T_m and T_n are relatively prime for $m \neq n$;

(c) $\displaystyle\sum_{k=1}^{n} \frac{1}{T_k} = 1 - \frac{1}{T_n(T_n - 1)}$;

(d) $T_n > n$.

5.3 Difference Methods

In this section we introduce some of the basic definitions and ideas of the subject called finite differences, and indicate how these ideas may be used to solve recurrence relations. We mentioned in Section 1.6 that difference methods are important in converting problems to be solved to forms that can be handled with the aid of a computer.

In Section 1.6 we defined the difference operator Δ by the equation

$$\Delta f(x) = f(x + 1) - f(x) \tag{1}$$

In general, Δ would be defined by $\Delta f(x) = f(x + h) - f(x)$. We have chosen $h = 1$. We call Δ the **difference operator**, h the **difference interval**, and Δf the **first difference of** f. Higher differences of f are defined iteratively by

$$\Delta^n f = \Delta(\Delta^{n-1} f), \qquad n = 2, 3, \ldots . \tag{2}$$

We define

$$\Delta^0 = I \tag{3}$$

where I is the **identity operator** defined by

$$If = f \tag{4}$$

for all f. Another operator, called the **translation operator**, E, is defined by

$$Ef(x) = f(x + 1). \tag{5}$$

In general it would be defined by $Ef(x) = f(x + h)$, but again we have chosen $h = 1$. Further powers of E are defined by

$$E^0 = I \tag{6}$$

and

$$E^n f(x) = E[E^{n-1} f(x)], \qquad n = 2, 3, \ldots . \tag{7}$$

EXAMPLE 1   For the function defined by $f(x) = x^3 + 3x - 4$ find
(a) $\Delta f(x)$, (b) $\Delta^2 f(x)$, (c) $\Delta^3 f(x)$, (d) $\Delta^4 f(x)$, (e) $Ef(x)$.

Solution

(a) $\Delta f(x) = f(x + 1) - f(x)$
$$= (x + 1)^3 + 3(x + 1) - 4 - (x^3 + 3x - 4)$$
$$= 3x^2 + 3x + 4.$$

(b) $\Delta^2 f(x) = \Delta(\Delta f(x))$
$$= 3(x + 1)^2 + 3(x + 1) + 4 - (3x^2 + 3x + 4)$$
$$= 6x + 6.$$

(c) $\Delta^3 f(x) = \Delta(\Delta^2 f(x))$
$$= 6(x + 1) + 6 - (6x + 6)$$
$$= 6.$$

(d) $\Delta^4 f(x) = \Delta(\Delta^3 f(x))$
$$= 6 - 6 = 0.$$

(e) $Ef(x) = (x + 1)^3 + 3(x + 1) - 4$
$$= x^3 + 3x^2 + 6x.$$

The following identities are listed and the reader may verify them. These identities, which are studied in finite differences, will not be used in this book.

For c a constant,

$$\Delta cf(x) = c\Delta f(x) \tag{8}$$

$$\Delta[f_1(x) + f_2(x)] = \Delta f_1(x) + \Delta f_2(x) \tag{9}$$

$$\Delta x^m = C(m, 1)x^{m-1} + C(m, 2)x^{m-2} + \cdots + C(m, r)x^{m-r} + \cdots + 1 \tag{10}$$

$$\Delta x^{(n)} = nx^{(n-1)} \tag{11}$$

$$\Delta^n x^{(n)} = n! \tag{12}$$

$$\Delta f(x) = Ef(x) - If(x) = (E - I)f(x)$$
$$\therefore \Delta \equiv E - I \tag{13}$$

$$E^n f(x) = f(x + n). \tag{14}$$

The difference operator is useful in evaluating certain sums. Consider the sum $\sum_{k=1}^{n} \Delta f(k)$. It represents

$$\Delta f(1) + \Delta f(2) + \cdots + \Delta f(n - 1) + \Delta f(n),$$

and we may write this sum as

$$\begin{aligned} &\quad f(2) - f(1) \\ + &\quad f(3) - f(2) \\ + &\quad f(4) - f(3) \\ &\qquad \vdots \\ + &\quad f(n) - f(n - 1) \\ + &f(n + 1) - f(n). \end{aligned}$$

Noting that pairs of terms cancel we see that

$$\sum_{k=1}^{n} \Delta f(k) = f(n + 1) - f(1). \tag{15}$$

EXAMPLE 2   Using the result (15) evaluate

$$\sum_{k=1}^{n} \frac{1}{k(k+1)}$$

Solution   By the method of partial fractions we may write

$$\frac{1}{k(k+1)} = \frac{1}{k} - \frac{1}{k+1}.$$

Hence

$$\sum_{k=1}^{n} \frac{1}{k(k+1)} = \sum_{k=1}^{n} - \left[\frac{1}{k+1} - \frac{1}{k}\right]$$

$$= -\sum_{k=1}^{n} \Delta f(k) \qquad \text{where} \quad f(k) = \frac{1}{k}$$

$$= -[f(n+1) - f(1)] \qquad \text{from (15)}$$

$$= -\left[\frac{1}{n+1} - 1\right]$$

$$= \frac{n}{n+1}.$$

By a **difference equation** we mean an equation involving the difference operator Δ, or equivalently the translation operator. For example, the difference equation $(E-2)f(n) = 0$ is equivalent to $(\Delta - 1)f(x) = 0$ since $E = \Delta + 1$ by (13). Another illustration is given in Example 3. The reader who is familiar with the methods of solution of differential equations will find that analagous methods are used to solve difference equations. For example, one method of solving a particular type of differential equation is to guess at a solution and then check out the guesswork. This method has its difference equation analogue as shown in Example 4.

EXAMPLE 3   Consider Equation (1) of Section 5.1, namely $S_{n+1} = 2S_n + 1$. If we write $f(n) = S_n + 1$ we see that

$$f(n+1) = S_{n+1} + 1$$
$$= (2S_n + 1) + 1$$
$$= 2(S_n + 1)$$
$$= 2f(n)$$

so that f satisfies

$$f(n+1) = 2f(n) \tag{16}$$

or, using (5),

$$Ef(n) = 2f(n).\tag{17}$$

This is sometimes written

$$(E - 2)f(n) = 0.\tag{18}$$

EXAMPLE 4   Solve the difference equation $(E - 2)f(n) = 0$.

Solution   We try to solve the given difference equation by guessing at the answer. Could f be a linear function, that is, could $f(n) = an + b$, where a and b are constants?

$$(E - 2)(an + b) = -an + a - b$$

and $-an + a - b = 0$ implies $n = (a - b)/a$, a constant, which, of course is a contradiction. Thus f is not a linear function. If we assume f is a quadratic function or any polynomial function we obtain a similar contradiction. The reader may check this.

Let us assume that f is an exponential function, that is, $f(n) = ca^n$, where c and a are real numbers. Then $Ef(n) = E(ca^n) = ca^{n+1}$ so that

$$(E - 2)ca^n = c(a^{n+1} - 2a^n)$$

and this equals zero if $a = 2$. Thus, $f(n) = ca^n$ is a possible solution of the difference equation if we choose $a = 2$. The value of c is determined by an initial condition. If $f(0) = 1$, then $c = 1$. The solution of the given equation $S_{n+1} = 2S_n + 1$ of Example 3 is then $S_n = f(n) - 1 = 2^n - 1$.

More formal methods of solving difference equations are available and are considered in a course on difference equations.

The equation in Example 3 can be solved directly without substitution. The reader is asked to do this in Problem 1 of Exercise 5.3.

EXERCISE 5.3

1.   (a)   Solve the equation $S_{n+1} = 2S_n + 1$ with the initial condition $S_1 = 3$ by the method of Example 3.
     (b)   Solve the same equation using difference methods but without using the substitution used in Example 3.

2.   For the functions in (i)–(v) find (a) $f(0)$, $f(1)$, $f(2)$ and (b) $\Delta f(0)$, $\Delta f(1)$, $\Delta f(2)$, $\Delta f(x)$:

   (i) $f(x) = 1$ (ii) $f(x) = x - 2$ (iii) $f(x) = x^2 + 1$
   (iv) $f(x) = x(x - 4)(x - 8)$ (v) $f(x) = 2^x$.

3. For each of the functions (i)–(v) of (2), find $\Delta^2 f(1)$ and $\Delta^2 f(x)$.

4. If $f(x) = 3^x$, determine a formula for $E^n f(x)$, $n = 0, 1, 2, \ldots$.

5. If $f(x)$ is a polynomial of degree n, find a formula for $\Delta^n f(x)$.

6. Prove that (a) $\Delta x^{(n)} = nx^{(n-1)}$, (b) $\Delta^n x^{(n)} = n!$.

7.* Justify the following symbolic expressions.

(a) $\Delta^n = \sum\limits_{k=0}^{n} \binom{n}{k}(-1)^{n-k}E^k$

(b) $E^n = \sum\limits_{k=0}^{n} \binom{n}{k}\Delta^k$

8.* Prove that $\Delta^n c^x = c^x \sum\limits_{k=0}^{n} \binom{n}{k}(-1)^{n-k}c^k$.

9. Evaluate the following sums using the relation (15) of this section

(a) $\sum\limits_{k=0}^{n} \dfrac{2}{(k+1)(k+2)}$

(b) $\sum\limits_{k=1}^{n} (2k+1)$

(c) $\sum\limits_{k=1}^{n} (3k^2 + 3k + 1)$

10. By noting that $\Delta k^{(3)} = 3k^{(2)}$, evaluate

$$\sum_{k=0}^{n} k(k-1).$$

5.4 A Fibonacci Sequence

The sequence

$$\{1, 1, 2, 3, 5, 8, 13, 21, \ldots\} \tag{1}$$

which satisfies the difference equation

$$F(n) = F(n-1) + F(n-2) \tag{2}$$

subject to the initial conditions $F(1) = F(2) = 1$, is called a Fibonacci sequence.

In this section we find a formula for the general term of the sequence (1) using the methods of Section 5.3.

The difference equation (2) can be written as

$$F(n + 2) - F(n + 1) - F(n) = 0$$

by replacing n by $n + 2$. In terms of the operator E this can be written in the form

$$(E^2 - E - 1)F(n) = 0.$$

We wish to find an expression for $F(n)$ that satisfies the equation. If we try linear or quadratic functions as in Example 3 of Section 5.3, we see that these do not yield the desired function. If we try $F(n) = a^n$ we obtain, successively,

$$(E^2 - E - 1)a^n = 0$$
$$a^{n+2} - a^{n+1} - a^n = 0$$
$$a^n(a^2 - a - 1) = 0$$
$$\therefore a^2 - a - 1 = 0$$
$$a = \frac{1 \pm \sqrt{5}}{2}.$$

The most general solution for $F(n)$ is then of the form

$$F(n) = c_1 \left(\frac{1 + \sqrt{5}}{2}\right)^n + c_2 \left(\frac{1 - \sqrt{5}}{2}\right)^n.$$

If we set $F(1) = F(2) = 1$, we obtain $c_1 = -c_2 = 1/\sqrt{5}$ and so

$$F(n) = \frac{1}{\sqrt{5}}\left[\left(\frac{1 + \sqrt{5}}{2}\right)^n - \left(\frac{1 - \sqrt{5}}{2}\right)^n\right].$$

This formula can be obtained by using generating functions, and this will be done in Section 6.3.

EXERCISE 5.4

1. Check that setting $F(1) = F(2) = 1$ yields $c_1 = -c_2 = 1/\sqrt{5}$ as claimed above.

2. Check that the initial conditions $F(0) = 0$, $F(1) = 1$ together with the difference equation (2) also yields the sequence (1). Obtain $c_1 = -c_2 = 1/\sqrt{5}$ much more easily by using these initial conditions.

5.5 A Summation Method

In Section 1.2 we solved the equation

$$f(n + 1) = f(n) + n + 1 \tag{1}$$

by summing the expression (1) for $n = 1, 2, \ldots, n$. This method will always work whenever we have an equation of the form $f(n + 1) = f(n) + a_n$, or equivalently, $\Delta f(n) = a_n$, and the sum $S_n = a_1 + a_2 + \cdots + a_n$ is known.

THEOREM   If $S_n = a_1 + a_2 + \cdots + a_n$, then the equation

$$f(n + 1) = f(n) + a_n$$

has solution $f(n + 1) = f(1) + S_n$.

This theorem is proved by summation as in Section 1.2.

EXERCISE 5.5

1. Solve the difference equations.

(a) $f(n + 1) = f(n) + n, f(1) = 2.$
(b) $f(n + 1) = f(n) + n^2, f(1) = 1.$
(c) $f(n + 1) = f(n) + n^3, f(1) = 1.$

2. Solve the following difference equations using the method of summation. [For (b)–(d) use the results of Example 2 and Problem 9 of Section 5.3.]

(a) $f(n + 1) = f(n) + n(n - 1), f(1) = 1.$
(b) $f(n + 1) = f(n) + \dfrac{1}{n(n + 1)}, f(1) = 1.$
(c) $f(n + 1) = f(n) + 2n + 1, f(1) = 1.$
(d) $f(n + 1) = f(n) + 3n^2 + 3n + 1, f(1) = 1.$
(e) $f(n + 1) = f(n) + n2^n, f(1) = 1.$

5.6 Chromatic Polynomials[1]

The problem of finding the chromatic polynomial of a map or of its corresponding graph was introduced in Sections 1.4 and 3.3. The method

---

[1] Much of the work of this section was derived from R. C. Read, An introduction to chromatic polynomials, *Journal of Combinatorial Theory*, **4** (1968) 52–71.

used in Section 1.4 for finding chromatic polynomials might be described as
a trial and error method. The method of Section 3.3 is more direct; however,
the number of cases to be considered in using this method is usually prohibi-
tive, even with the use of a computer, except in very simple situations. In
this section we present an iterative method for finding the chromatic polynomial $P(G, \lambda)$ of a graph G. This is the best method which has so far been
developed for finding chromatic polynomials.

Recall that a graph is a geometric object consisting of points (nodes,
vertices), pairs of which may be joined by line segments (edges). Two vertices
are said to be **adjacent** if they are joined by an edge. In the problem of coloring
a graph we require that adjacent vertices be colored differently. In Figure 5.1,

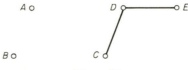

Figure 5.1

for example, there are no restrictions on the colors that can be used for A
and B; C and D cannot be colored alike, nor can D and E. However, C and
E can be colored alike.

The **empty graph** with n vertices is defined to be the graph with n vertices
and no edges. The **complete graph** on n vertices is defined to be the graph on n
vertices in which each pair of vertices is joined by an edge. A graph G is
connected if there is a path (collection of adjoining edges) joining every pair
of distinct vertices in G. A component G_k of G is a maximally connected subgraph of G; that is, G_k is not properly contained in any other connected subgraph of G. The graph (a) in Figure 5.2 is the empty graph on four vertices;

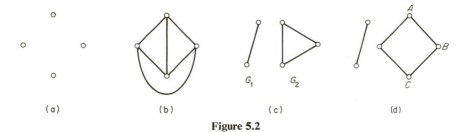

Figure 5.2

graph (b) is the complete graph on four vertices; graph (c) consists of two
components G_1 and G_2. In (d), the subgraph consisting of vertices A, B,
and C and the edges joining these vertices is not a component of the graph
because it is contained in a "larger" connected subgraph.

The reader is asked, in Problem 1 of Exercise 5.6, to prove the following three theorems.

THEOREM 1   The chromatic polynomial of the empty graph on n vertices is λ^n.

THEOREM 2   The chromatic polynomial of the complete graph on n vertices is $\lambda^{(n)}$.

THEOREM 3   If G consists of two components G_1 and G_2, then

$$P(G, \lambda) = P(G_1, \lambda) \cdot P(G_2, \lambda)$$

where $P(G, \lambda)$ denotes the chromatic polynomial of the graph G.

EXAMPLE 1   Find chromatic polynomials of graphs (a),(b),(c), in Figure 5.2.

Solution   (a)   The chromatic polynomial of the graph in (a) is λ^4 from Theorem 1.

(b)   The chromatic polynomial of the graph in (b) is

$$\lambda^{(4)} = \lambda(\lambda - 1)(\lambda - 2)(\lambda - 3)$$

by Theorem 2.

(c)   By Theorem 3, the chromatic polynomial of the graph in (c) is

$$P(G_1, \lambda) \cdot P(G_2, \lambda) = \lambda^{(2)} \cdot \lambda^{(3)} = \lambda^2(\lambda - 1)^2(\lambda - 2).$$

Let G denote a graph whose vertices are to be colored. Let e denote any edge of G and define two associated graphs G_e' and G_e'' of G relative to edge e as follows:

(i) G_e' is the graph derived from G by deleting the edge $e = (A, B)$ but retaining its end points A and B.

(ii) G_e'' is the graph derived from G_e' by identifying the vertices that were the end points of e, that is, A and B.

EXAMPLE 2   Find G_e' and G_e'' if

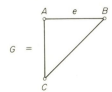

Solution

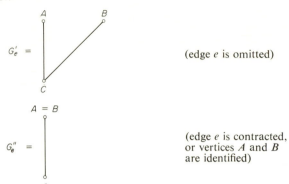

G'_e =   (edge *e* is omitted)

G''_e =   (edge *e* is contracted,
or vertices *A* and *B*
are identified)

We should really write

G''_e =

showing two edges joining *C* and $A = B$. However, as far as colorings are concerned, these two graphs are equivalent. The *number* of edges joining two vertices is immaterial; the important fact is whether or not the vertices are joined by an edge.

EXAMPLE 3   Find G'_e and G''_e if

G =

Solution

G'_e = G''_e =

The extra dotted edge from *C* to $A = D$ in G''_e is omitted.

EXAMPLE 4   Find G_e' and G_e'' if

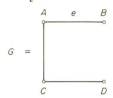

Solution

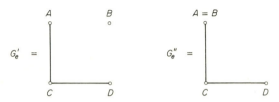

The following theorem is basic in the present theory.

THEOREM 4   Let $P(G, \lambda)$ be the chromatic polynomial of a graph G with edge e. Then

$$P(G, \lambda) = P(G_e', \lambda) - P(G_e'', \lambda).$$

Proof The number of ways of coloring G_e' includes all the ways of coloring G *plus* the colorings in which the vertices of edge e have the same color. But this latter number is the same as the number of ways of coloring G_e''.

A study of the following example should reinforce the reader's understanding of the preceding theorem.

EXAMPLE 5   Use Theorem 4 to find the chromatic polynomial of the graph

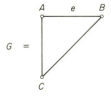

Solution We know from Theorem 2 that $P(G, \lambda) = \lambda^{(3)}$. We can verify this by using Theorem 4 as follows.

$P(G_e', \lambda) = \lambda(\lambda - 1)^2$ since C can be colored in λ ways and then A and B each in $\lambda - 1$ ways. By Theorem 2,

$$P(G_e'', \lambda) = \lambda^{(2)} = \lambda(\lambda - 1).$$

Then

$$\begin{aligned}
P(G_e', \lambda) - P(G_e'', \lambda) &= \lambda(\lambda - 1)^2 - \lambda(\lambda - 1) \\
&= \lambda(\lambda - 1)[\lambda - 1 - 1] \\
&= \lambda(\lambda - 1)(\lambda - 2) \\
&= \lambda^{(3)} = P(G, \lambda).
\end{aligned}$$

Note that every coloring of G is also a coloring of G_e'. However, every coloring of G has to have A and B colored differently, whereas A and B can be colored the same in G_e'. This means that the number of colorings of G_e' exceeds the number of colorings of G by the number of colorings in which A and B are colored the same. But the colorings of G_e'' will be the colorings of G in which A and B are colored alike.

EXAMPLE 6   Use Theorem 4 to find the chromatic polynomial of the graph

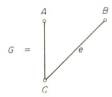

Solution

$$G_e' = \qquad\qquad\qquad G_e'' =$$

Now, by Theorem 3,

$$P(G_e', \lambda) = \lambda P(G_e'', \lambda)$$

Then

$$\begin{aligned}
P(G, \lambda) &= P(G_e', \lambda) - P(G_e'', \lambda) \\
&= (\lambda - 1)P(G_e'', \lambda) \\
&= (\lambda - 1)\lambda^{(2)} \qquad \text{(by Theorem 2)} \\
&= \lambda(\lambda - 1)^2.
\end{aligned}$$

In finding the chromatic polynomial of a given graph, the recurrence relation in Theorem 4 may be applied interatively until the required chromatic polynomial is expressed in terms of known chromatic polynomials. To shorten the work, we normally use a *formal* sum or difference of graphs to represent the sum or difference of the corresponding chromatic polynomials of those graphs; that is, we use the graphs themselves to represent their chromatic polynomials. Thus in Example 6 we would write

to indicate that

$$P(G, \lambda) = P(G_e', \lambda) - P(G_e'', \lambda).$$

Example 5 could then be completed as follows.

$$= (\lambda - 2)\lambda^{(2)}$$
$$= \lambda^{(3)}.$$

EXAMPLE 7   Find the chromatic polynomial of the graph

$$G \ = \ $$

Solution

$$= (\lambda - 2) \ \triangleleft$$

$$= (\lambda - 2)\lambda^{(3)}$$
$$= \lambda(\lambda - 1)(\lambda - 2)^2.$$

Alternative Solution

$$= (\lambda - 2) \left\{ \ \right\} - \ \triangle$$

$$= (\lambda - 2) \left\{ \ \ \circ \ - \ \right\} - \ \triangle$$

$$= \left\{ (\lambda - 2)(\lambda - 1) \ \ \right\} - \ \triangle$$

$$= (\lambda - 2)(\lambda - 1)\lambda^{(2)} - \lambda^{(3)}$$
$$= \lambda(\lambda - 1)(\lambda - 2)^2 \qquad \text{as before.}$$

The alternate solution is used to indicate that the choice of edges e in the iteration could help to reduce the work. The first solution is a little more direct.

The iterative process outlined above could be termed a decreasing one in the sense that we are decreasing the number of edges one at a time and could eventually end up expressing the chromatic polynomials of the given graph in terms of chromatic polynomials of empty graphs. This is shown in Example 8.

EXAMPLE 8   Express the chromatic polynomial of the graph

in terms of chromatic polynomials of empty graphs.

Solution

$$= (\lambda - 2)(\lambda^2 - \lambda).$$

We may also use an "increasing" interative process in the sense that we *add* edges and express the given graph in terms of complete graphs whose chromatic polynomials are known. To do this we rewrite the formula of Theorem 4 as

$$P(G_e', \lambda) = P(G, \lambda) + P(G_e'', \lambda)$$

We formulate this result in the following theorem.

THEOREM 5   Let $G_e{}^+$ be defined as the graph G with an edge e added joining two nonadjacent vertices A and B, and let $G_e{}^{++}$ be G with vertices A and B identified. Then

$$P(G,\ \lambda) = P(G_e{}^+,\ \lambda) + P(G_e{}^{++},\ \lambda).$$

EXAMPLE 9   Rework Example 7 using the "increasing" iterative process established in Theorem 5.

Solution

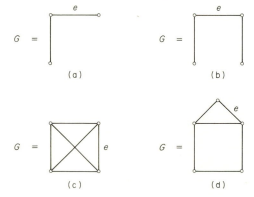

$$= \lambda^{(4)} + \lambda^{(3)} \qquad\qquad\qquad \text{(by Theorem 2)}$$
$$= \lambda(\lambda - 1)(\lambda - 2)(\lambda - 3) + \lambda(\lambda - 1)(\lambda - 2)$$
$$= \lambda(\lambda - 1)(\lambda - 2)^2, \qquad\qquad \text{(as before)}.$$

It can be seen that by using the method of Theorem 5 the chromatic polynomial of any graph can be expressed as a sum of factorials $\lambda^{(n)}$, and hence is indeed a polynomial. When a chromatic polynomial is expressed in this way, we say that it is in factorial form. If the factorial form is known, the polynomial itself is readily found. This is facilitated by the use of Stirling numbers. Tables of Stirling numbers, which allow us to express the $\lambda^{(n)}$ as polynomials in λ, are discussed in Section 13.4.

EXERCISE 5.6

1. Prove Theorems 1, 2, and 3 of this section.

2. Find $G_e{}'$ and $G_e{}''$ for the following graphs G and indicated edges e.

3. Find $G_e{}^+$ and $G_e{}^{++}$ for the following graphs G using the indicated non-adjacent vertices.

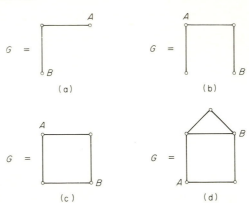

4. Find chromatic polynomials of the following graphs using the two methods indicated in Theorems 4 and 5.

5. In how many ways can the following graphs be colored with the given number of colors?

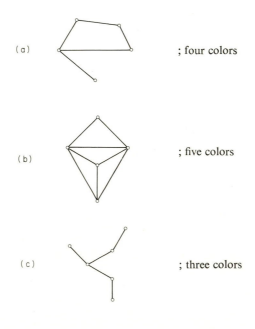

(a)   ; four colors

(b)   ; five colors

(c)   ; three colors

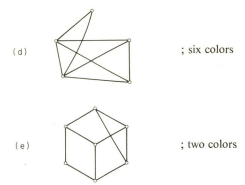

(d)   ; six colors

(e)   ; two colors

6.   In how many ways can the faces of a cube be colored with four colors so that adjacent faces are colored differently?

7.   In how many ways can the walls and roof of a barn (with an ordinary peak roof) be painted so that no two adjacent faces are the same color if three colors are available?

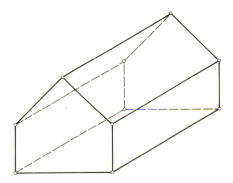

8.   In how many ways can the faces of an octahedron be colored with two colors so that faces with a common edge have different colors?

9.   In how many ways can the faces of a hexagonal pyramid, not including the base, be colored with four colors so that adjacent faces receive different colors?

10.* Develop an iterative method for expressing the chromatic polynomial of a given graph as the sum of chromatic polynomials of trees.
     Hint: At each iteration, eliminate an edge which is in a circuit (that is, on a closed path).

11.[c]* Write a computer program for finding the chromatic polynomial of a graph by employing the technique of Theorem 4, expressing the given graph in terms of empty graphs. Check your program with Example 8.

12.[c]* Write a computer program for finding the chromatic polynomial of a given graph using the technique of Theorem 5. Check your program with Example 7.

13.[c]* Write a computer subroutine for determining whether or not a given edge is on a circuit of a graph. Hence modify your program in Problem 11 to express the chromatic polynomial of a given graph as a sum of chromatic polynomials of trees (see Problem 10).

Generating Functions

6.1 Some Simple Examples

The reader may check that

$$x^2 = 1(1) - 2(1 + x) + 1(1 + x)^2.$$

We say that x^2 determines or generates the sequence $\{1, -2, 1\}$ of coefficients relative to the sequence of functions $\{1, (1 + x), (1 + x)^2\}$. Similarly, $16x^2 - 3$ determines the sequence $\{3, 1, 8\}$ relative to $\{1 - x, 2 + 3x, 2x^2 - 1\}$ since

$$16x^2 - 3 = 3(1 - x) + 1(2 + 3x) + 8(2x^2 - 1).$$

If, for a function f,

$$f(x) = 0(1) + 1 \sin x + 2 \sin 2x + \cdots + n \sin nx + \cdots$$

then f determines or generates the infinite sequence of real numbers

$$\{0, 1, 2, \ldots, n, \ldots\}$$

relative to the infinite sequence of functions

$$\{1, \sin x, \sin 2x, \ldots, \sin nx, \ldots\}.$$

109

In general, if f is a function such that

$$f(x) = a_0 \, u_0(x) + a_1 \, u_1(x) + a_2 \, u_2(x) + \cdots + a_n \, u_n(x) + \cdots,$$

where the a_i are real numbers, we say that f generates or is the generating function of the sequence of real numbers

$$\{a_0, a_1, a_2, \ldots, a_n, \ldots\}$$

with respect to the sequence of functions

$$\{u_0, u_1, u_2, \ldots, u_n, \ldots\}.$$

For example, if

$$\phi(x) = a_0 + a_1 \frac{x}{1!} + a_2 \frac{x^2}{2!} + \cdots + a_n \frac{x^n}{n!} + \cdots \tag{1}$$

we call ϕ the **exponential** generating function of the sequence $\{a_n\}$. The name stems from the fact that the exponential generating function of the sequence $\{1, 1, 1, \ldots\}$ is e^x.

In our work we will restrict our attention to generating functions with respect to the sequence of functions

$$\{1, x, x^2, \ldots, x^n, \ldots\}$$

and with this convention we will say that if

$$f(x) = a_0 + a_1 x + a_2 x^2 + \cdots + a_n x^n + \cdots, \tag{2}$$

then f is the **generating function** of the sequence $\{a_n\}$. For example, since

$$\frac{1}{1 - 2x} = 1 + 2x + 2^2 x^2 + \cdots + 2^n x^n + \cdots, \tag{3}$$

as can be verified using the binomial theorem, then the function f with $f(x) = 1/(1 - 2x)$ is the generating function of the sequence $\{1, 2, 2^2, \ldots, 2^n, \ldots\}$. (We usually say simply that $1/(1 - 2x)$ is the generating function.)

In considering a sum such as in (2) or (3) we are not interested in questions of convergence of the infinite series but rather in the way in which the function on the left is represented as a power series in x, that is, we are interested in the sequence of coefficients. We say that the power series we consider are **formal power series** in x and x is a **formal mark** rather than a variable. The theory of formal power series allows us to ignore questions of convergence. Thus we write

$$\frac{1}{1 - 2x} = 1 + 2x + 4x^2 + \cdots$$

even though the statement would be nonsense if we were to set $x = 1$, for example. (For a development of the theory of formal power series, see the article "Formal Power Series" by Ivan Niven, *American Mathematical Monthly*, **76** (1969), 871–889.)

One of the most familiar examples of a generating function is given by the binomial theorem which asserts that $(1 + x)^n$ is the generating function for the sequence

$$\left\{ \binom{n}{0}, \binom{n}{1}, \binom{n}{2}, \cdots, \binom{n}{r}, \cdots \right\}$$

of binomial coefficients. Further simple examples are listed below.

EXAMPLE 1   The constant function 1 is the generating function for the sequence $\{1, 0, 0, \ldots\}$.

EXAMPLE 2   The identity function defined by $f(x) = x$ is the generating function for the sequence $\{0, 1, 0, 0, \ldots\}$.

EXAMPLE 3   Generalizing Examples 1 and 2, the function defined by $f(x) = x^j$ is the generating function for the sequence

$$\{\delta_{0j}, \delta_{1j}, \delta_{2j}, \ldots, \delta_{ij}, \ldots\}$$

where δ_{ij} is the Kronecker delta, which is a symbol defined as follows: $\delta_{ij} = 1$ for $i = j$, $\delta_{ij} = 0$ if $i \neq j$.

EXAMPLE 4   Since

$$\frac{1}{1 - x} = 1 + x + x^2 + \cdots + x^n + \cdots,$$

the function defined by $f(x) = 1/(1 - x)$ is the generating function for the sequence $\{1, 1, 1, \ldots\}$.

EXAMPLE 5   Since

$$(1 - x)^{-2} = \sum_{r=0}^{\infty} \binom{r + 1}{r} x^r$$

$$= 1 + 2x + 3x^2 + \cdots + (r + 1) x^r + \cdots$$

the function defined by $f(x) = (1 - x)^{-2}$ is the generating function for the sequence

$$\{1, 2, 3, \ldots, n + 1, \ldots\}.$$

If derivatives of $f(x)$ can be computed, then we may be able to use the calculus to find the terms a_n of the sequence generated by the function f, according to the following theorem.

THEOREM   If f is the generating function of the sequence $\{a_n\}$ and if f and its derivatives of all orders exist for $x = 0$, then

$$a_n = \frac{f^{(n)}(0)}{n!}. \qquad (4)$$

To obtain this result we find successive derivatives of f by differentiating the right-hand side of (2) repeatedly and then setting $x = 0$ in f and its derivatives and solving for the coefficients a_n. This theorem is impractical for finding a_n except in a small number of cases.

We should point out that in some applications of probability (for example, random walks along the x axis), it is convenient to allow negative powers as well as positive powers. In such cases the generating function has the form

$$g(x) = \sum_{n=-\infty}^{\infty} a_n x^n.$$

Generating functions can also be defined when more than one parameter is involved. For example, the generating function

$$h(x, y) = \sum_{i,j} a_{ij} x^i y^j$$

corresponds to the "sequence" $\{a_{ij}\}$.

EXERCISE 6.1

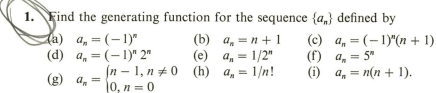

1. Find the generating function for the sequence $\{a_n\}$ defined by

(a) $a_n = (-1)^n$ (b) $a_n = n + 1$ (c) $a_n = (-1)^n(n + 1)$
(d) $a_n = (-1)^n 2^n$ (e) $a_n = 1/2^n$ (f) $a_n = 5^n$
(g) $a_n = \begin{cases} n - 1, n \neq 0 \\ 0, n = 0 \end{cases}$ (h) $a_n = 1/n!$ (i) $a_n = n(n + 1)$.

2. Use the theorem of this section to evaluate the binomial coefficients from the corresponding generating function $f(x) = (1 + x)^n$.

3. Use the theorem of this section to determine the sequences generated by

(a) e^x (b) $\sin x$ (c) $\cos x$.

6.2 The Solution of Difference Equations by Means of Generating Functions

In this section we will give some examples of how generating functions can be used to find a formula for $f(x)$ when the coefficients in the expansion of $f(x)$ satisfy a difference equation and suitable initial conditions. A formula for the general term a_n of the sequence generated by f can then be obtained by finding the coefficient of x^n in the power series expansion for $f(x)$. In Section 1.1 we used this method to solve the difference equation $t_n = 2t_{n-1}$ with $t_0 = 1$; the reader should review this example.

EXAMPLE 1   Use a generating function to solve the difference equation $s_{k+1} = s_k + 2$ with $s_0 = 1$.

Solution   Let

$$f(x) = s_0 + s_1 x + s_2 x^2 + \cdots + s_n x^n + s_{n+1} x^{n+1} + \cdots$$

so that f generates the sequence determined by the given difference equation. As we saw in Section 1.1, our modus operandi is to obtain expressions of the form $(s_{n+1} - s_n - 2)x^{n+1}$ which all have the value 0 since $s_{n+1} = s_n + 2$ from the given difference equation. This is done as follows.

$$
\begin{aligned}
f(x) = \ & s_0 + s_1 x + s_2 x^2 + \cdots + s_n x^n + s_{n+1} x^{n+1} + \cdots \\
-x f(x) = \ & \quad - s_0 x - s_1 x^2 - \cdots - s_{n-1} x^n - s_n x^{n+1} - \cdots \\
-2\left(\frac{1}{1-x}\right) = \ & -2 - 2x - 2x^2 - \cdots - 2x^n - 2x^{n-1} - \cdots
\end{aligned}
$$

$$
\begin{aligned}
\therefore (1-x)f(x) - \frac{2}{1-x} &= (s_0 - 2) + (s_1 - s_0 - 2)x + (s_2 - s_1 - 2)x^2 \\
&\quad + \cdots + (s_n - s_{n-1} - 2)x^n + (s_{n+1} - s_n - 2)x^{n+1} + \cdots \\
&= s_0 - 2 = -1
\end{aligned}
$$

since all remaining coefficients on the right side are zero. Thus

$$(1-x)f(x) = \frac{2}{1-x} - 1 = \frac{1+x}{1-x}$$

and

$$
\begin{aligned}
f(x) &= \frac{1+x}{(1-x)^2} \\
&= (1-x)^{-2} + x(1-x)^{-2} \\
&= [1 + 2x + 3x^2 + \cdots + (n+1)x^n + \cdots] \\
&\quad + [x + 2x^2 + 3x^3 + \cdots + nx^n + \cdots]
\end{aligned}
$$

(from Example 5 of Section 6.1)

$$= 1 + 3x + 5x^2 + \cdots + (2n+1)x^n + \cdots.$$

It follows that s_n, the coefficient of x^n in the expansion of $f(x)$, must be $2n + 1$ and the solution of the given difference equation is $s_n = 2n + 1$.

EXAMPLE 2   Solve the difference equation $a_n = 2a_{n-1} - a_{n-2}$ with $a_0 = 0$ and $a_1 = 1$.

Solution   Let

$$f(x) = a_0 + a_1 x \quad + a_2 x^2 + \cdots + \qquad a_n x^n + \cdots$$
$$-2xf(x) = \qquad -2a_0 x - 2 a_1 x^2 - \cdots - 2a_{n-1} x^n + \cdots$$
$$x^2 f(x) = \qquad\qquad a_0 x^2 + \cdots + a_{n-2} x^n + \cdots$$

$$\therefore (1 - 2x + x^2)f(x) = a_0 + (a_1 - 2a_0)x + (a_2 - 2a_1 + a_0)x^2$$
$$+ \cdots + (a_n - 2a_{n-1} + a_{n-2})x^n + \cdots$$
$$= a_0 + (a_1 - 2a_0)x$$
$$= x.$$

Then

$$f(x) = \frac{x}{(1-x)^2} = x(1-x)^{-2}$$
$$= x(1 + 2x + 3x^2 + \cdots + (n+1)x^n + \cdots)$$

(from Example 5 of Section 6.1)

$$= x + 2x^2 + 3x^3 + \cdots + nx^n + \cdots$$

so that $a_n = n$ is the required solution.

EXAMPLE 3   Solve the difference equation $s_{k+1} = s_k + k + 1$ with $s_0 = 0$. (See Example 1 of Section 5.2.)

Solution   Here we require the series

$$\frac{x}{(1-x)^2} = x(1 + 2x + 3x^2 + \cdots + (n-1)x^{n-2} + nx^{n-1} + \cdots)$$
$$= x + 2x^2 + 3x^3 + \cdots + (n-1)x^{n-1} + nx^n + \cdots.$$

Let

$$f(x) = s_0 + s_1 x + s_2 x^2 + \cdots + \quad s_n x^n + \cdots$$
$$-xf(x) = \quad - s_0 x - s_1 x^2 - \cdots - s_{n-1} x^n - \cdots$$

$$-\frac{x}{(1-x)^2} = \quad - \quad x - \quad 2x^2 - \cdots - \quad nx^n - \cdots$$

$$(1-x)f(x) - \frac{x}{(1-x)^2} = s_0 \text{ (since all other coefficients are zero)}$$

$$= 0$$

$$\therefore f(x) = x(1-x)^{-3} = x \sum_{r=0}^{\infty} \binom{3+r-1}{r} x^r$$

$$= x \sum_{r=0}^{\infty} \binom{r+2}{r} x^r$$

$$\therefore s_n = \binom{n-1+2}{n-1} = \binom{n+1}{n-1} = \frac{1}{2} n(n+1),$$

as we found in Example 1 of Section 5.2.

EXERCISE 6.2 

Use generating functions to solve the following difference equations.

1. $s_{n+1} = s_n + 3$ with $s_0 = 1$.

2. $a_{n+1} = a_n + n$ with $a_0 = 1$.

3. $t_{n+1} = t_n + \frac{1}{2}(n+1)(n+2)$ with $t_0 = 0$ (cf. Example 2, Section 5.2).

4. $a_{n+1} = a_n + 2n + 1$, $a_0 = 0$.

5. $a_{n+1} = a_n + 4 - n$, $a_0 = 0$.

6. $a_{n+1} = a_n + 2^n$, $a_0 = 0$.

7. $y_{k+1} = Ay_k + B$ where A and B are constants, $A \neq 1$, and $y_0 = 1$.

8. Show that $x(1 + x)(1 - x)^{-3}$ is the generating function for the sequence $\{1^2, 2^2, \ldots, n^2, \ldots\}$ and use this information to solve the difference equation $f(n + 1) = f(n) + (n + 1)^2$, $f(0) = 0$, of Example 3, Section 5.2.

9. If $A(x) = \sum_{n=0}^{\infty} a_n x^n$, what is the generating function for

  (a) $\{b_n\}$ where $b_n = a_{n-1}$ for $n \neq 0$, $b_0 = 0$
  (b) $\{c_n\}$ where $c_n = a_{n+1}$?

6.3 Some Combinatorial Identities

  Generating functions can be used to prove certain combinatorial identities. In the simplest case the method used is based on the fact that

$$\sum_{n=0}^{\infty} a_n x^n = \sum_{n=0}^{\infty} b_n x^n$$

if and only if $a_n = b_n$, $n = 0, 1, 2, \ldots$. (This is the method of "comparing like coefficients" used in Chapter 2.)

EXAMPLE 1   Prove that

$$\sum_{k=0}^{\infty} (-1)^{r+k} \binom{n}{k} \binom{k}{r} = \delta_{nr}.$$

  Solution   Consider n fixed. The generating function of the right-hand side is

$$\sum_{r=0}^{\infty} \delta_{nr} x^r$$

which equals x^n since $\delta_{nr} = 0$ if $r \neq n$ and $\delta_{nn} = 1$. The generating function of the left side is

$$\sum_{r=0}^{\infty} \left\{ \sum_{k=0}^{n} (-1)^{r+k} \binom{n}{k} \binom{k}{r} \right\} x^r = \sum_{k=0}^{n} (-1)^k \binom{n}{k} \left[\sum_{r=0}^{\infty} \binom{k}{r} (-1)^r x^r \right]$$

$$= \sum_{k=0}^{n} (-1)^k \binom{n}{k} (1 - x)^k$$

$$= [1 - (1 - x)]^n$$

$$= x^n.$$

The generating functions are equal, hence, by "comparing like coefficients," the identity is proved.

1	0	0	0	0	0
1	1	0	0	0	0
1	2	1	0	0	0
1	3	3	1	0	0
1	4	6	4	1	0
1	5	10	10	5	1

Figure 6.1

EXAMPLE 2   Prove that the sum of the entries in the $(k + 1)$st column of the Pascal triangle down to the $(p + 1)$st row is

$$A_{k+1} = \binom{0}{k} + \binom{1}{k} + \cdots + \binom{p}{k} = \sum_{n=0}^{p} \binom{n}{k} = \binom{p+1}{k+1}.$$

Solution   To illustrate this result, consider the entries in the fourth column down to the sixth row in the Pascal triangle (Figure 6.1). We have supplied zeros for missing entries in the array. The required sum is

$$\binom{0}{3} + \binom{1}{3} + \binom{2}{3} + \binom{3}{3} + \binom{4}{3} + \binom{5}{3} = 15$$

and $\binom{6}{4} = 15$.

$$f(x) = \sum_{k=0}^{\infty} A_{k+1} x^k = \sum_{k=0}^{\infty} \sum_{n=0}^{p} \binom{n}{k} x^k$$

$$= \sum_{n=0}^{p} \left[\sum_{k=0}^{\infty} \binom{n}{k} x^k \right]$$

$$= \sum_{n=0}^{p} (1 + x)^n$$

$$= \frac{(1 + x)^{p+1} - 1}{(1 + x) - 1}$$

$$= \frac{1}{x} [(1 + x)^{p+1} - 1]$$

$$= \frac{1}{x} \left[\sum_{r=0}^{p+1} \binom{p+1}{r} x^r - 1 \right]$$

$$= \sum_{r=1}^{p+1} \binom{p+1}{r} x^{r-1}$$

$$= \sum_{k=0}^{p} \binom{p+1}{k+1} x^k, \qquad \text{setting } k = r - 1.$$

Equating coefficients yields

$$A_{k+1} = \sum_{n=0}^{p} \binom{n}{k} = \binom{p+1}{k+1},$$

as required.

EXERCISE 6.3

1.* Prove, using generating functions, that

$$\sum_{j=0}^{r} \binom{n}{r-j}(-1)^j = \binom{n-1}{r}.$$

2. If $f(x)$ is the generating function for $\{a_n\}$ and $g(x)$ is the generating function for $\{b_n\}$, for what sequence is $f(x) + g(x)$ the generating function? For what sequence is $f(x)\,g(x)$ the generating function?

3.* Using (2) and generating functions, prove that

$$\sum_{r=0}^{k} \binom{a}{r}\binom{b}{k-r} = \binom{a+b}{k}. \qquad$$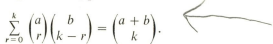

4.* Using (2) and generating functions, prove that

$$\binom{n+1}{r} = \binom{n}{r} + \binom{n}{r-1}.$$

6.4 Additional Examples

In this section we show how generating functions can be used to obtain two results that have already been obtained using other methods. The first is the finding of a formula for the general term of the Fibonacci sequence. This was found in Section 5.4 by guessing at a solution to the difference equation $(E^2 - E - 1)a^n = 0$. The second is the formula for the number of combinations, m at a time, of r distinct objects each available in unlimited supply. This formula was established in the theorem of Section 4.3; the proof was related to the number of solutions of a linear equation with unit coefficients.

EXAMPLE 1   Use a generating function to solve the difference equation

$$F(n) = F(n-1) + F(n-2), \qquad F(1) = F(2) = 1,$$

which corresponds to a Fibonacci sequence.

Solution   Let
$$f(x) = F(1) + F(2)x + F(3)x^2 + \cdots + F(n)x^{n-1} \quad + \cdots$$
$$xf(x) = \qquad\quad F(1)x + F(2)x^2 + \cdots + F(n-1)x^{n-1} + \cdots$$
$$x^2 f(x) = \qquad\qquad\qquad F(1)x^2 + \cdots + F(n-2)x^{n-1} + \cdots$$

$$(1 - x - x^2)f(x) = F(1) + [F(2) - F(1)]x$$
$$= 1.$$
$$\therefore f(x) = \frac{1}{1 - x - x^2} = \frac{1}{(1 - \alpha x)(1 - \beta x)}.$$

where $\alpha = (1 + \sqrt{5})/2$ and $\beta = (1 - \sqrt{5})/2$.
By the method of partial fractions

$$\frac{1}{(1 - \alpha x)(1 - \beta x)} = \frac{A}{1 - \alpha x} + \frac{B}{1 - \beta x} \quad \text{where} \quad A = \frac{\alpha}{\alpha - \beta}, \quad B = \frac{-\beta}{\alpha - \beta}.$$

Thus
$$f(x) = A(1 - \alpha x)^{-1} + B(1 - \beta x)^{-1}$$
$$= A \sum_{n=0}^{\infty} \alpha^n x^n + B \sum_{n=0}^{\infty} \beta^n x^n.$$

Now $F(n)$ is the coefficient of x^{n-1} so that
$$F(n) = A\alpha^{n-1} + B\beta^{n-1}$$
$$= \frac{\alpha^n - \beta^n}{\alpha - \beta}$$
$$= \frac{1}{\sqrt{5}} \left[\left(\frac{1 + \sqrt{5}}{2}\right)^n - \left(\frac{1 - \sqrt{5}}{2}\right)^n \right]$$

as found in Section 5.4.

EXAMPLE 2   Use a generating function to show that the number of combinations of r distinct objects taken m at a time, with unlimited repetitions allowed, is $C(r + m - 1, m)$.

Solution   Let the number of combinations of r objects taken m at a time, with unlimited repetitions allowed, be $f(r, m)$. Let the r objects be $x_1, x_2, \ldots, x_r$. Then any combination of m of these either includes a specific object x_i or it does not. This fact yields the recurrence relation

$$f(r, m) = f(r, m - 1) + f(r - 1, m) \tag{1}$$

where $f(r - 1, m)$ is the number of combinations not including the specific object x_i and $f(r, m - 1)$ is the number of combinations including x_i. Now consider the generating function

$$F_r(x) = \sum_{m=0}^{\infty} f(r, m)x^m$$

$$= \sum_{m=0}^{\infty} f(r-1, m)x^m + x \sum_{m-1=0}^{\infty} f(r, m-1)x^{m-1} \qquad \text{[using (1)]}$$

$$= F_{r-1}(x) + xF_r(x).$$

Thus

$$(1 - x)F_r(x) = F_{r-1}(x)$$

and

$$F_r(x) = \frac{F_{r-1}(x)}{1 - x}.$$

At this point we note that $F_0(x) = 1$ implies that

$$F_1(x) = \frac{1}{1 - x}$$

and we may prove by induction that $F_r(x) = (1 - x)^{-r}$. Then

$$F_r(x) = (1 - x)^{-r} = \sum_{m=0}^{\infty} \binom{m + r - 1}{m} x^r$$

so that

$$f(r, m) = \binom{m + r - 1}{m},$$

as required.

EXERCISE 6.4

1. Use generating functions to find a formula for the general term of the Fibonacci sequence defined by

$$a_0 = 2, \ a_1 = 5, \ a_{n+2} = a_{n+1} + a_n$$

for $n = 0, 1, 2, \ldots$.

2. Use generating functions to solve

$$y_{k+2} - 2y_{k+1} + y_k = 2^k, \qquad y_0 = y_1 = 1.$$

3.* An ordering of the integers $1, 2, \ldots, n$ is called *acceptable* if, except for the first integer, k appears in the ordering only if $k - 1$ or $k + 1$ has appeared before it.

(a)   Obtain a recurrence relation for $f(n, r)$, the number of acceptable orderings of $1, 2, \ldots, n$ which begin with r.
(b)   Use generating functions to find $f(n, r)$.
(c)   Find $f(n)$, the total number of acceptable orderings.
(d)   Check (c) for $n = 4$.
(e)   Can you obtain $f(n, r)$ more directly?

4.* Use generating functions to solve

$$a_n = \sum_{k=1}^{n-1} a_k a_{n-k}, \qquad a_1 = a_2 = 1.$$

6.5 Derivatives and Differential Equations

Differential calculus can be employed effectively to find generating functions. A few examples are presented to illustrate some of the ideas involved. Those readers unfamiliar with the calculus should omit this section.

EXAMPLE 1   In section 6.1 it was pointed out that the function defined by

$$f(x) = (1 + x)^\alpha, \tag{1}$$

α not necessarily integral, is the generating function for the sequence $\{\binom{\alpha}{i}\}$, that is,

$$(1 + x)^\alpha = \sum_i \binom{\alpha}{i} x^i. \tag{2}$$

By differentiating the identity (2) successively, new identities are obtained that yield generating functions for related sequences.

$$f'(x) = \alpha(1 + x)^{\alpha-1} = \sum_i i \binom{\alpha}{i} x^{i-1},$$

that is,

$$\alpha(1 + x)^{\alpha-1} = \sum_i (i + 1) \binom{\alpha}{i + 1} x^i \tag{3}$$

showing that the sequence $\{(i + 1) \binom{\alpha}{i+1}\}$ has the generating function $\alpha(1 + \alpha)^{\alpha-1}$. Differentiating again yields

$$\alpha^{(2)}(1+x)^{\alpha-2} = \sum_i (i+2)^{(2)} \binom{\alpha}{i+2} x^i,$$

and in general

$$\alpha^{(k)}(1+x)^{\alpha-k} = \sum_i (i+k)^{(k)} \binom{\alpha}{i+k} x^i. \tag{4}$$

From this latter identity it follows that the generating function corresponding to the sequence

$$\left\{ \frac{(i+k)^{(k)}}{\alpha^{(k)}} \binom{\alpha}{i+k} \right\} \quad \text{is} \quad (1+x)^{\alpha-k}.$$

If we set $x = 1$ in (1), (2), (3), or (4) we obtain interesting identities involving the binomial coefficients.

In the following example we use a generating function in the solution of a differential equation.

EXAMPLE 1   Solve the differential equation

$$f'(x) = f(x) \qquad \text{with} \quad f(0) = 1.$$

Solution   Let

$$f(x) = a_0 + a_1 x + a_2 x^2 + \cdots + a_n x^n + \cdots,$$

so that f is the generating function of the sequence $\{a_n\}$. Then

$$f'(x) = a_1 + 2a_2 x + 3a_3 x^2 + \cdots + na_n x^{n-1} + (n+1)a_{n+1}x^n + \cdots$$

so that f' is the generating function of the sequence $\{(n+1)a_{n+1}\}$. The differential equation relating f and f' will always produce a difference equation relating the coefficients a_n. In this case we are given that $f'(x) = f(x)$ with $f(0) = 1$. This means that

$$(n+1)a_{n+1} = a_n \qquad \text{for} \quad n \geq 0$$

with $a_0 = f(0) = 1$. We may solve this difference equation by inspection to obtain

$$a_n = \frac{1}{n!}.$$

Thus

$$f(x) = \sum_{n=0}^{\infty} \frac{x^n}{n!} = e^x$$

is the required solution of the given differential equation.

In the next example we use derivatives to solve the given difference equation.

EXAMPLE 2   Solve the difference equation

$$(n + 1)a_{n+1} = a_n + \frac{1}{n!}, \qquad a_0 = 1.$$

Solution   Let f be the generating function corresponding to $\{a_n\}$. Then

$$\sum_{n=0}^{\infty} (n + 1)a_{n+1}x^n = \sum_{n=0}^{\infty} \left(a_n + \frac{1}{n!}\right)x^n$$

yields

$$f'(x) = f(x) + e^x \qquad \text{with} \quad f(0) = a_0 = 1. \tag{5}$$

We may rewrite (5) as

$$e^{-x}f'(x) - e^{-x}f(x) = 1$$

so that

$$[e^{-x}f(x)]' = 1$$

and

$$e^{-x}f(x) = f(0) + x = 1 + x.$$

We may now write

$$f(x) = e^x + xe^x$$

$$= \sum_{n=0}^{\infty} \frac{x^n}{n!} + \sum_{n=1}^{\infty} \frac{x^n}{(n-1)!}$$

so that

$$a_n = \frac{1}{n!} + \frac{1}{(n-1)!}$$

is the solution to the given difference equation.

In the final example of this section we use the calculus to produce the generating function of a given sequence.

EXAMPLE 3   Find the generating function for the sequence $\{a_n\}$ with $a_n = \binom{2n}{n}$, $n \neq 0$, $a_0 = 1$.

Solution

$$a_{n+1} = \binom{2n+2}{n+1} = \frac{(2n+2)(2n+1)(2n)!}{(n+1)^2 n! \, n!}$$

$$= \frac{4n+2}{n+1}\binom{2n}{n} = \frac{4n+2}{n+1}a_n$$

so that the terms of the given sequence obey the difference equation

$$(n + 1)a_{n+1} = (4n + 2)a_n.$$

Now let

$$f(x) = \sum_{n=0}^{\infty} a_n x^n.$$

Then

$$\sum_{n=0}^{\infty} (n + 1)a_{n+1}x^n = 4 \sum_{n=0}^{\infty} na_n x^n + 2 \sum_{n=0}^{\infty} a_n x^n$$

so that

$$f'(x) = 4xf'(x) + 2f(x)$$

or

$$\frac{f'(x)}{f(x)} = \frac{2}{1 - 4x}.$$

Integrating, we obtain

$$\log f(x) = -\tfrac{1}{2} \log(1 - 4x) + C.$$

When $x = 0$, $f(x) = a_0 = 1$; thus $C = 0$ and

$$\log f(x) = \log(1 - 4x)^{-1/2}.$$

Thus $f(x) = (1 - 4x)^{1/2}$ is the required generating function.

EXERCISE 6.5

1. (a) Prove $\displaystyle\sum_{r=0}^{n} r\binom{n}{r} = n2^{n-1}.$

    (b) Prove $\displaystyle\sum_{r=0}^{n} r^2\binom{n}{r} = n(n + 1)2^{n-2}.$

2. Evaluate

    (a) $\displaystyle\sum_{r=0}^{n} \frac{1}{r + 1}\binom{n}{r}$

    (b) $\displaystyle\sum_{k=0}^{n} (-1)^k(k + 1)\binom{n}{k}.$

3. Evaluate $\displaystyle\sum_{n=1}^{\infty} \frac{n^3}{n!}$.

4. Solve $(n + 1)f_{n+1} = (2n + 4)f_n$, $f_0 = 1$.

5. Verify the result of Example 3 by expanding $(1 - 4x)^{-1/2}$ by the binomial theorem.

6.* Using the result of Example 3, prove that

$$\sum_{r=0}^{n} \binom{2r}{r}\binom{2n - 2r}{n - r} = 4^n.$$

Hint: How is the generating function of 4^n related to that of $\binom{2n}{n}$?

7. Define a_0 to be 1 and for $n \geq 1$, let a_n be the number of symmetric $n \times n$ matrices with exactly one 1 and $n - 1$ zeros in each row and each column. Prove that

$$a_{n+1} = a_n + na_{n-1}.$$

8.* In (7), let $f(x) = \displaystyle\sum_{n=0}^{\infty} a_n \frac{x^n}{n!}$. Prove that $f(x) = e^{x + \frac{1}{2}x^2}$.

II

EXISTENCE

Existence proofs play an important role in modern mathematics, especially in mathematical systems which are defined axiomatically. In Chapter 7 a number of methods for proving existence are discussed and illustrated by examples. Most existence proofs involve combinations or modifications of these basic methods. The remaining chapters of Part II present topics in combinatorics in which existence plays a dominant role.

Chapter 8 illustrates the use of existence in problems involving the geometry of the plane. Chapter 9 contains a discussion of polyhedra; in particular the derivation of the platonic solids. The problem of coloring maps on a sphere and related topics are considered in Chapter 10; this chapter includes additional work on chromatic polynomials. Chapter 11 contains a collection of topics that are needed to generalize ordinary analytic geometry of the plane to finite geometries.

Some Methods of Proof

The idea of existence was discussed in Section 1.5. There are two classes of methods of proof, direct and indirect. In direct methods, existence is proved by construction and nonexistence is proved by the method of exhaustion, that is, by showing that none of the possible objects is the required object. These methods are illustrated in Sections 7.1 and 7.2. Indirect methods include the application of Dirichlet's drawer principle which was illustrated in Section 1.5 and the method of averaging. These methods are illustrated in Section 7.3. Nonexistence can be proved indirectly by contradiction, as illustrated in Section 7.4. Further examples of all these methods of proof are given in Chapters 8, 9, 10, and 11.

7.1 Existence by Construction

If the desired object can be constructed, then its existence is assured.

EXAMPLE 1 Prove that there exists a map with eight regions which can be colored in two colors.

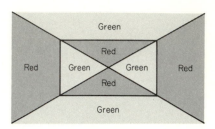

Figure 7.1

Solution   The proof consists of demonstrating a map with the required properties. The map shown in Figure 7.1 is such a map.

The subject of coloring maps is discussed further in Section 10.1. Some people think that every possible map that can be drawn in a plane can be colored with at most four colors, but no one has proved this conjecture. If someone (and he would be famous!) could construct a map that could *not* be colored with at most four colors, that is, one that would require at least five colors, then the "four color conjecture" would be disproved.

EXAMPLE 2   Prove that a matrix of order three can be constructed such that each row and each column is a 3-arrangement or permutation of the symbols 1, 2, and 3.

Solution   Such a matrix can be found by trial and error. We can set the symbols 1, 2, and 3 in the first row and first column of the matrix as shown in Figure 7.2a. Now we have two choices, either 1 or 3, for the term a_{22} (the term in the second row and second column). Suppose we choose $a_{22} = 1$ (Figure 7.2b); a_{23} must equal 3 and then the third column cannot be a 3-arrangement of the given symbols. Hence we must choose $a_{22} = 3$ and the remaining entries fall into place (Figure 7.2c). The reader can find other such

Figure 7.2

matrices by trial and error. Such matrices are examples of **latin squares**, matrices of order n in which each row and each column is an n-arrangement of the symbols 1, 2, ..., n. In Section 16.2 we give a method for constructing Latin squares and also show that for some n it is possible to construct a pair of **orthogonal latin squares**, that is, a pair of latin squares $A = (a_{ij})$ and $B = (b_{ij})$ which are such that all n^2 ordered pairs (a_{ij}, b_{ij}) for $i, j = 1, 2, \ldots, n$, are different. A pair of orthogonal latin squares is given in Example 3.

EXAMPLE 3   Show that the following latin squares are orthogonal.

$$A = (a_{ij}) = \begin{bmatrix} 1 & 2 & 3 \\ 2 & 3 & 1 \\ 3 & 1 & 2 \end{bmatrix}, \qquad B = (b_{ij}) = \begin{bmatrix} 1 & 2 & 3 \\ 3 & 1 & 2 \\ 2 & 3 & 1 \end{bmatrix}.$$

Solution   We note by inspection that the given matrices are latin squares. Now check the nine ordered pairs (a_{ij}, b_{ij}) for $i, j = 1, 2, 3$. These are the following:

$$(1, 1), \ (2, 2), \ (3, 3),$$

$$(2, 3), \ (3, 1), \ (1, 2),$$

$$(3, 2), \ (1, 3), \ (2, 1).$$

The nine ordered pairs are all different and so the given latin squares are orthogonal.

EXAMPLE 4   Prove that there exists a finite number system in which operations of addition and multiplication are defined and all the ordinary rules of arithmetic hold.

Solution   Consider the set {*ODD, EVEN*} with addition and multiplication defined as in Table 7.1. The object *ODD* behaves like 1 under multiplication and *EVEN* behaves like 0 under addition and multiplication. Subtraction is defined in the usual way; thus

$$EVEN - EVEN = EVEN, \qquad EVEN - ODD = ODD$$
$$ODD - EVEN = ODD, \qquad ODD - ODD = EVEN.$$

TABLE 7.1

+	EVEN	ODD		×	EVEN	ODD
EVEN	EVEN	ODD		EVEN	EVEN	EVEN
ODD	ODD	EVEN		ODD	EVEN	ODD

Division, $x \div y$, is defined only for $y \neq EVEN$:

$$EVEN \div ODD = EVEN, \qquad ODD \div ODD = ODD.$$

The reader may verify that all the ordinary rules of arithmetic hold.

The number system defined in Example 4 is a finite field and is sometimes referred to as the Galois field on two elements. A further discussion is given in Section 11.1.

EXERCISE 7.1

1. (a) Show that there exists a map containing four regions each of which has three edges.

    (b) Show that there exists a map containing six regions each of which has four edges.

2. Show that there exists a finite number system containing three elements 0, 1, and 2, for which addition and multiplication can be defined in such a way that all the ordinary rules of arithmetic hold.

3. Construct a pair of orthogonal 3×3 latin squares that are different from the ones given in Example 3.

7.2 The Method of Exhaustion

We may know that objects of a given type exist and we may wish to know whether one or more of these objects has a certain property. This can be checked by the method of exhaustion, that is, by an examination of the objects, one by one. Existence of an object with a given property will be proved as soon as an object with the given property has been found. Nonexistence will have been proven after all the objects have been examined and none found with the desired property. In many combinatorial problems the number of objects involved is large and a computer is a useful tool in the application of this method.

EXAMPLE 1 Are there any integral roots of the equation

$$f(x) = x^6 - 4x^3 + x^2 - 5x + 8 = 0?$$

Solution From the theory of equations we know that any integral roots of the given equation must be divisors of eight and so we may check the existence of integral roots by evaluating $f(m)$ for all integral divisors m of eight.

$$f(1) = 1, \quad f(-1) = 19, \quad f(2) = 34, \quad f(-2) = 118,$$
$$f(4) = 3844, \quad f(-4) = 4396, \quad f(8) = 260{,}128, \quad f(-8) = 264{,}304.$$

We see that there are no integral roots of the given equation. Of course, it should be pointed out that we did not have to calculate each of the above $f(m)$ but simply to satisfy ourselves that $f(m) \neq 0$ in each case.

EXAMPLE 2   Is it possible to label the vertices and edges of a triangle with the integers $1, 2, \ldots, 6$ in such a way that the sums along each side of the triangle are the same ?

Solution   We shall use the method of trial and error. A sequence of possible steps is indicated in Figure 7.3. In (a) we label one of the vertices

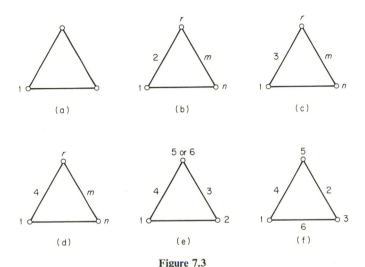

Figure 7.3

with 1. An adjoining edge must then be labeled 2, 3, 4, 5, or 6. In (b) we assume that it is 2. But then $m + n = 3$ (since $m + n + r = 3 + r$) which is impossible because both m and n have to be one of 3, 4, 5, or 6. In (c) we assume that an adjoining edge is labeled 3. But then $m + n = 4$, which is impossible since m and n are either 2, 4, 5, or 6 and $m \neq n$. In (d) we assume that an adjoining edge is labeled 4. Then $m + n = 5$, and m and n can be either 2 or 3. Now, in (e), if $m = 3$ and $n = 2$, then $r = 5$ or 6, neither of which is possible since

$$1 + 4 + 5 = 2 + 3 + 5 \neq 1 + 6 + 2$$

and

$$1 + 4 + 6 = 2 + 3 + 6 \neq 1 + 5 + 2.$$

Finally, in (f), $m = 2$ and $n = 3$ implies $r = 5$ or 6 and $r = 5$ gives a required solution. Choosing $r = 6$ does not give a solution.

The concepts used in the next example are discussed further in Section 7.4 and Chapter 9.

EXAMPLE 3   Prove that it is impossible to label the twelve faces of a regular dodecahedron using all of the integers 1, 2, . . . , 12, in such a way that the numbers on adjacent faces always differ by more than two.

Solution   We show a planar representation of the dodecahedron in Figure 7.4 and we have labeled the twelve faces $a, b, c, \ldots, l$ for reference. All

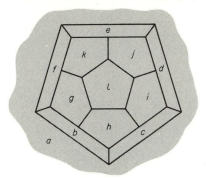

Figure 7.4

of the faces are equivalent so we may label the face a with the integer 1 without loss of generality. Now since the numbers with which each pair of faces are labeled must differ by more than 2, we cannot use 2 or 3 in faces b, c, d, e, or f. There are only two essentially different faces that can be labeled 2, namely face l and face j (or i or h or g or k). If, as in (a) of Figure 7.5, we label face l as 2, we see that we have no face to label 3; it cannot be used for any of the remaining ten faces without violating the restriction that the numbers on adjacent faces must always differ by more than two.

In (b) of Figure 7.5 we then label face j with 2; then we cannot use 3 for faces b, c, d, e, f, i, k, or l and so 3 can be used only for faces g or h and it does not matter which one it is used for. Suppose then that we use 3 to label

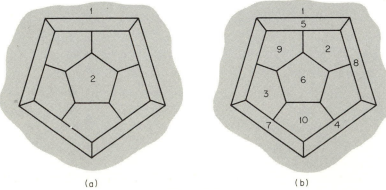

(a)   (b)

Figure 7.5

face *g*. This means that 4 cannot be used for faces *b, f, h, k, l,* or *d, e, i,* and so must be used for face *c*. Continuing, we find that 5 must be used for face *e*, then 6 for face *l*, 7 for face *b*, 8 for face *d*, 9 for face *k*, and 10 for face *h*. Now there are only two faces, *i* and *f*, unlabeled, and two integers 11 and 12 remaining, and we find that 11 cannot be used for either face *i* of *f* without violating the laid down condition, although 12 can be used to label face *f*. We conclude that it is impossible to label the faces of the dodecahedron as asserted in the statement of the problem.

E X E R C I S E   7.2

1. Show that a solution exists in Example 2 in which an edge adjoining vertex 1 is labeled 5.

2. Is it possible to label the triangle as in Example 2 but in which 1 is used to label an edge?

3. Show that it is possible to label the vertices and edges of a square with the integers 1, 2, ..., 8 (no repetitions) in such a way that the sum along each side is the same.

4. Show that it is possible to label the vertices and sides and diagonal of the square

with the integers 1, 2, ..., 9 (no repetitions) in such a way that the sums along the sides and diagonal are the same.

5. Show that it is possible to label the sides and diagonals of the square

with the digits 1, 2, ..., 9 (one digit repeated once) in such a way that the sums along the sides and diagonals are the same.

6. Is it possible to label the eight faces of the graph

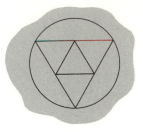

with the integers 1, 2, ..., 8 in such a way that the numbers on adjacent faces always differ by more than 2?

7. Is it possible to label the seven faces of the graph

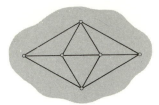

with the digits 1, 2, ..., 7 (no repetitions) in such a way that numbers on adjacent faces differ (a) by at least 2, (b) by more than 2?

8.* Is it possible to label the vertices and edges of the tetrahedron with the digits 1, 2, ..., 10 (no repetitions) in such a way that the sum along each edge is the same?

7.3 The Dirichlet Drawer Principle

The Dirichlet drawer principle, also known as the pigeonhole principle, was used in Section 1.5. The principle may be stated as follows.

DIRICHLET DRAWER PRINCIPLE

If $k + 1$ objects are placed in k drawers, at least one drawer will contain two or more objects.

EXAMPLE 1   (a)   In a group of eight people, at least two will have their birthday on the same day of the week.

(b)   In a group of thirteen people, at least two will have their birthday in the same month.

(c)   In a group of 367 people, at least two will have their birthday on the same day and month.

This principle only determines existence in Example 1 and not which specific individuals have the required birthdays.

Extensions of the Dirichlet drawer principle are indicated in Exercise 7.3.

THE METHOD OF AVERAGING

The method of averaging, which is closely connected to the Dirichlet drawer principle, is illustrated in the following examples.

Example 2 involves a geometrical object called a **polyhedron**, a geometrical object consisting of vertices, edges, and faces which are plane *n*-gons. Examples of *convex polyhedra* include the cube and the tetrahedron. Convex polyhedra are discussed in Chapter 9. In Example 2 we assume a very important formula, Euler's polyhedron formula, which states that, for a convex polyhedron,

$$V - E + F = 2,$$

where V is the number of vertices, E the number of edges, and F the number of faces of the polyhedron. Examples of the application of this formula for various polyhedra and a further discussion of the formula are given in Section 9.1.

EXAMPLE 2   If a given convex polyhedron has six vertices and twelve edges, prove that every face is a triangle.

Solution   We are given that $V = 6$ and $E = 12$ so that $F = 8$ by Euler's theorem. If q is the average number of edges per face, then, counting all the edges on all the faces, since each edge is counted twice we must have $qF = 2E$. Thus

$$q = \frac{2E}{F} = \frac{24}{8} = 3.$$

But no face can have *fewer* than three edges. Hence, since the average number of edges per face is three, no face can have *more* than three edges, and so each face has exactly three edges and is a triangle.

EXAMPLE 3   A circle is divided into 200 congruent sectors and 100 of the sectors are colored red and the other 100 are colored blue. A smaller circle is also divided into 200 congruent sectors with 100 colored red and 100 blue. The smaller circle is placed concentrically on the larger one. Prove that no matter how the 100 red sectors are chosen for each circle, the smaller circle can be rotated so that at least 100 sectors of the two circles match in color.

Solution   As the smaller circle rotates through 360°, a given sector S_1 of the smaller circle matches in color with 100 of the sectors of the larger circle, and similarly for sectors $S_2, \ldots, S_{200}$ of the smaller circle. Hence there are altogether 20,000 color matches, which means an average of 100 matches per position. Since the average must be attained or exceeded there must be some position with at least 100 sectors matching in color.

EXERCISE 7.3

1.   Prove that if $2k + 1$ objects are placed in k drawers, at least one drawer will contain three or more objects.

2.   Generalize Problem 1 to $mk + 1$ objects placed in k drawers.

3.   At registration time, 750 freshmen were required to select exactly five courses from a total of ten. Show that among all such possible combinations of courses, there was at least one not selected by more than two students.

7.4 The Method of Contradiction

It is sometimes possible to show the nonexistence of an object with a given property by proving that the existence of such an object would imply a property which is known to be false.

The example of this section involves the question of the existence of a particular finite field. We mentioned in Section 7.1 that a finite field is an algebraic system with a finite number of elements which can be added and multiplied with all the usual arithmetic rules holding. A further discussion is given in Section 11.1.

EXAMPLE   Show that there does not exist a finite field $(F, +, \times)$ where F is a set containing exactly six elements and $+$ and $\times$ are "addition" and "multiplication."

Solution   We know that F contains the two distinct elements 0 and 1. Consider the sequence

$$0, 1, 2 = 1 + 1, 3 = 1 + 1 + 1, 4 = 1 + 1 + 1 + 1, \ldots .$$

Since we have postulated a finite number of elements in our field, there must be two elements

$$m = (\text{sum of } m \text{ 1's}) \qquad \text{and} \qquad n = (\text{sum of } n \text{ 1's})$$

with integer $m >$ integer n but with $m = n$ in the finite field. Now subtract n 1's from both sides of the equation $m = n$ to obtain

$$k = m - n = \text{sum of } (m - n) \text{ 1's} = 0.$$

We call k the **characteristic** of the finite field if k is the minimum number with this property. (For the fields of rational and real numbers, k is said to be 0; for the finite field of Example 4 of Section 7.1, $k = 2$.)

Now $k \neq 1$, since $1 \neq 0$. Assume $k = 2$. Let a denote an element of F with $a \neq 0, 1$. The reader should check that $x + ya$ yields four different elements of F for $x, y = 0, 1$. Then there is an element b different from any of these four elements. But this would yield eight distinct elements of the form $x + ya + zb$ for $x, y, z = 0, 1$. Since this is impossible (we have only six elements) it follows that $k \neq 2$.

A similar argument shows that $k \neq 3$, 4, or 5 and so k must equal 6. But then

$$2 \times 3 = (1 + 1)(1 + 1 + 1) = 1 + 1 + 1 + 1 + 1 + 1 = 6 \times 1 = 0$$

and this cannot occur in a field, since in a field the product of two elements is zero if and only if one of the elements is zero (one of the ordinary rules of arithmetic).

Since we have now shown that there is no possible choice for k, we may conclude that a field containing six elements does not exist.

EXERCISE 7.4

1. Show in the example of this section that $k = 3$, 4, and 5 lead to contradictions.

2. Show that finite fields with (a) ten elements, (b) 55 elements do not exist.

3.* Does there exist a finite field with four elements? (See Section 11.1.)

Geometry of the Plane

This chapter contains several examples illustrating existence which are not normally included in courses in the geometry of the plane. We discuss convex sets, tilings of the rectangle, tessellations of the plane, and equivalence classes of demi-dominoes. Convex sets are important in optimization theory, particularly in linear programming. Although the tiling of rectangles with noncongruent squares appears to be devoid of application, it is closely related to ideas in electrical engineering. Tessellations are interesting plane figures often mentioned in books on mathematical recreations. The idea of equivalence classes is a useful algebraic concept which has important implications in geometry.

8.1 Convex Sets

A subset of the Euclidean plane is **convex** if any line segment joining two points of the set lies entirely in the set. This definition can be extended to n dimensions. In Figure 8.1 we show a plane figure that is convex and one that is not convex. The **convex hull** $H(S)$ of a set S is the "smallest" convex set containing all the points of *S*, that is, the intersection of all convex sets containing all the points of *S*. Some examples are shown in Figure 8.2.

We prove the following result concerning convex sets.

THEOREM   Let five points be chosen in the plane with no three collinear. Then there exists a subset of four of the points which are the vertices of a convex quadrilateral.

Proof If the convex hull of the five points is a pentagon or a quadrilateral, the proof is immediate (Figure 8.3a, b). If the convex hull is a triangle (Figure 8.3c), then two of the five given points, *A* and *B*, will be

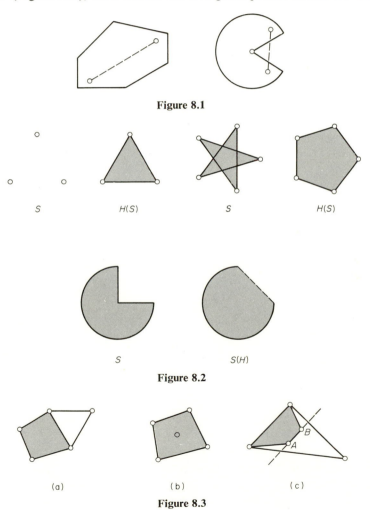

Figure 8.1

S   *H(S)*   *S*   *H(S)*

S   *S(H)*

Figure 8.2

(a)   (b)   (c)

Figure 8.3

inside the triangle formed by the other three and the two vertices of the triangle on the same side of the line AB, along with vertices A and B, form the required convex quadrilateral. This proves the theorem.

The result of this theorem suggests a more general problem. What is the maximum number of points that can be arranged in the plane with no three collinear so that no n of them form a convex n-gon? If we let this maximum number be $f(n)$, then clearly $f(3) = 2$. Also $f(4) = 4$. This result follows from the theorem of this section and the fact that it is possible to locate four points in the plane with no three collinear so that no convex quadrilateral is formed.

It was conjectured that $f(n)$ as defined above satisfies the inequality

$$2^{n-2} \leq f(n) \leq \binom{2n-4}{n-2}$$

and this conjecture has been proved recently by Kalbfleisch and Stanton, who also proved that $f(5) = 8$ in a paper in the *Proceedings of the 1970 Louisiana Conference on Combinatorics*. It has also been conjectured that $f(n) = 2^{n-2}$. This conjecture is true for $n = 3, 4,$ and 5, but it has not yet been established for other values of n.

EXERCISE 8.1

1. Determine which of the following sets are convex:

    (a) the points interior to an obtuse triangle;
    (b) the points on the circumference of a circle;
    (c) the points interior to an ellipse.

2. Find the convex hull of each of the sets described in Exercise 1.

3. Find the convex hull of the set of points on (a) a parabola, (b) a hyperbola.

8.2 Tiling a Rectangle

Is it possible to tile a given rectangle, that is, to partition it into congruent squares? into incongruent squares (squares all of different sizes)? Is it possible to partition a square into congruent or incongruent squares? Is it possible to partition an equilateral triangle into incongruent equilateral triangles? The answers to these questions are discussed in this section. It is clear that it is possible to partition some rectangles into congruent squares and equilateral triangles into congruent equilateral triangles but Tutte[1] proved in 1948 that it

---

[1] W. T. Tutte, The dissection of equilateral triangles into equilateral triangles, *Proceedings Cambridge Philosophical Society*, **44** (1948), 463–482.

is impossible to partition an equilateral triangle into incongruent equilateral triangles. In this section, we show that some rectangles can be partitioned into incongruent squares, and we prove that it is impossible to partition a rectangular box into a finite number of incongruent cubes.

THEOREM 1   An $a \times b$ rectangle can be partitioned into congruent squares if and only if a/b is rational.

Proof   If a partition of an $a \times b$ rectangle into congruent squares of side λ exists, then $a = m\lambda$ and $b = n\lambda$, where m and n are integers, so that

$$\frac{a}{b} = \frac{m\lambda}{n\lambda} = \frac{m}{n},$$

a rational number. Conversely, if a/b is rational, then $a/b = m/n$, with m and n integers, so that $a/m = b/n = \lambda$ and $a = m\lambda$, $b = n\lambda$. Thus we can divide the given rectangle into mn squares of side λ.

The existence of a partition of a rectangle into incongruent squares is demonstrated below. We first prove some theorems that will be useful in the discussion.

THEOREM 2   If a rectangle can be partitioned into incongruent squares, the smallest square cannot be on the boundary of the rectangle.

Proof   Assume that a rectangle has been partitioned into incongruent squares and that the smallest square is (i) in a corner, (ii) on a side of the given rectangle. Let the side of the smallest square be x (Figure 8.4). Then, since it

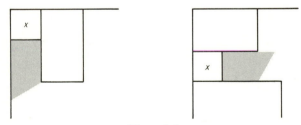

Figure 8.4

must be bounded by larger squares, in each case we obtain a shaded area which leads to a contradiction. The contradiction arises since the shaded area can only be filled by squares of side $< x$, or a square of side equal to x, and x is assumed to be the smallest of a set of incongruent squares.

THEOREM 3   If a rectangle can be partitioned into incongruent squares, the smallest square must be surrounded by exactly four larger squares.

Proof　If a rectangle can be partitioned into incongruent squares, the smallest must lie in the interior by Theorem 2. The smallest square, of side x, can meet a larger square, of side y, in four possible positions as shown in Figure 8.5. The first position, shown in (a), leads to the situation shown in Figure 8.6 in which the shaded area can only be filled by squares of side $\leq x$, a contradiction. Similarly the situations in (b) and (c) lead to contradictions. This leaves only the fourth possibility, in (d), which leads to the situation in Figure 8.7 which is the one described in the statement of the theorem. This proves the theorem.

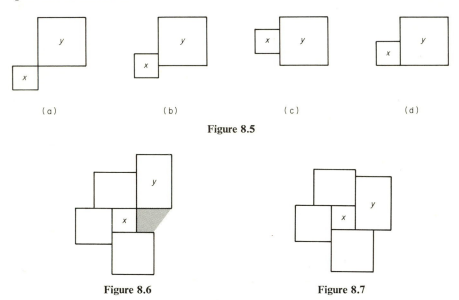

(a)　　　　　　　　　(b)　　　　　　　　　(c)　　　　　　　　　(d)

Figure 8.5

Figure 8.6　　　　　　　　　　　　　**Figure 8.7**

THEOREM 4　It is impossible to partition a rectangle into n incongruent squares if $n \leq 5$.

Proof　From Theorem 3, n must be at least five since the smallest square must be bounded by four larger ones. Assume $n = 5$. Then the partition must be as in Figure 8.8 (the figure is not drawn to scale but merely indicates the

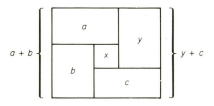

Figure 8.8

partitioning). Let the smallest square have side x and a second one have side y. Then the remaining squares have sides a, b, and c where $a = y - x$, $b = y - 2x$, and $c = x + y$. Then the left side of the rectangle has length $2y - 3x$ and the right side has length $2y + x$. Obviously $2y - 3x = 2y + x$ which implies that $x = 0$. We conclude that no partition of the rectangle into five incongruent squares exists.

It is also possible to show that partitions of the rectangle into six, seven, and eight incongruent squares are not possible (Problem 4 of Exercise 8.2).

EXAMPLE   Show that the partitioning indicated in Figure 8.9 leads to a partition of a rectangle into nine incongruent squares.

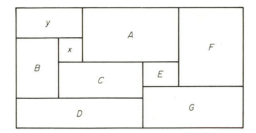

Figure 8.9

Solution   Label the "squares" as shown and let the labels indicate the lengths of sides of the corresponding squares. Then $A = x + y$, $B = y - x$, $C = y - 2x$, $D = 2y - 3x$, $E = 4x$, $F = 5x + y$, and $G = 9x + y$. Equating the lengths of the left and right sides of the rectangle yields $4y - 4x = 14x + 2y$ or $y = 9x$. The dimensions of the rectangle are then found to be $32x$ and $33x$ and we conclude that for any real number x, such a rectangle can be partitioned into incongruent squares. If $x = 1$, the squares have sides 1, 4, 7, 8, 9, 10, 14, 15, and 18. The reader should draw such a rectangle and its partitioning to scale using squared paper.

The result of the above example proves Theorem 5.

THEOREM 5   There exist rectangles which can be partitioned into incongruent squares.

It was shown by J. Wilson[2] that a square can be partitioned into 25 incongruent squares.

[2] J. Wilson, A Method for Finding Simple Perfect Square Squarings, 1964, Ph.D. thesis, University of Waterloo.

Finally we prove the following theorem.

THEOREM 6   It is impossible to partition a rectangular box into a finite number of incongruent cubes.

Proof   Suppose such a partitioning were possible. Then the rectangular bottom of the box would be partitioned into congruent squares by the incongruent cubes touching the bottom of the box. The smallest square x does not touch the boundary of the rectangular bottom by Theorem 2, so that the cubes on the bottom which surround the cube C_1 on the smallest square x must extend above C_1. The top of C_1 must then be covered by cubes and these cubes partition the top of C_1 into incongruent squares. This leads to a cube C_2 on C_1 which is surrounded by larger cubes. Continuing in this way, we are led to an infinite sequence $C_1, C_2, \ldots$ of incongruent cubes in the rectangular box, contrary to our assumption of a finite number. This contradiction means that a partitioning such as we assumed is not possible, and so our theorem is proved.

EXERCISE 8.2

1. Is it possible to partition an $a \times b$ rectangle into congruent squares if

    (a) $a = 3\frac{1}{2}, b = 4\frac{1}{6}$
    (b) $a = 3\sqrt{5}, b = 7\sqrt{5}$
    (c) $a = \sqrt{5}, b = \sqrt{15}$?

2. If a/b is rational, determine the smallest number of congruent squares required to partition an $a \times b$ rectangle. What is the size of each square?

3. Show that an equilateral triangle can be partitioned into (*a*) four, (*b*) nine congruent equilateral triangles.

4. Prove that it is impossible to partition a rectangle into (a) six, (b) seven, (c) eight incongruent squares.

5. Show that the situations shown in (b) and (c) of Figure 8.5 lead to contradictions in Theorem 3.

6. Does the partitioning indicated in Figure 8.10 lead to a partitioning of a rectangle into incongruent squares?

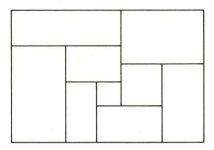

Figure 8.10

7.* Find, if possible, a partitioning of a rectangle into nine incongruent squares which is essentially different from the one in the example.

8.* Is it possible to have a partitioning of a rectangle into incongruent squares in which four squares have a common vertex?

9. Find conditions on a, b, and c so that a rectangular box of dimensions a, b, and c can be partitioned into congruent cubes. Construct the corresponding partition.

8.3 Tessellations of the Plane

The plane can be covered with congruent squares of any size as illustrated in Figure 8.11. Such a cover of the plane by congruent regular n-gons is called a **regular tessellation of the plane**. Regular tessellations of the plane by triangles and hexagons are illustrated in Figures 8.12 and 8.13.

THEOREM 1   There exist regular tessellations of the plane by n-gons for $n = 3, 4, 6$. No other regular tessellations of the plane exist.

Proof   The first part of the theorem is proved by Figures 8.11, 8.12, and 8.13. Suppose a tessellation by n-gons is possible. Let m denote the number

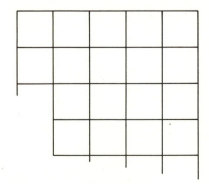

Figure 8.11

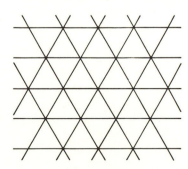

Figure 8.12

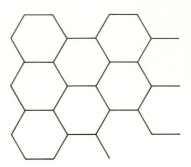

Figure 8.13

of n-gons at every vertex. Since the angle of an n-gon is $\left(1 - \dfrac{2}{n}\right)\pi$ it follows that

$$m\left(1 - \frac{2}{n}\right)\pi = 2\pi, \text{ that is,}$$

$$\frac{1}{n} + \frac{1}{m} = \frac{1}{2}.$$

The only solutions of this equation in positive integers are

$$
\begin{aligned}
n &= 3, & m &= 6 \\
n &= 4, & m &= 4 \\
n &= 6, & m &= 3
\end{aligned}
$$

corresponding to triangles, squares, and hexagons respectively, (as shown in Figure 8.11, 8.12, and 8.13 respectively). Hence no other tessellations exist.

Tessellations are also possible in which different polygons are used with the same set of polygons at each vertex (See Figures 8.14–8.21). These are **homogeneous** tessellations. The determination of all homogeneous tessellations is left as an exercise.

EXERCISE 8.3

1.   (a)   Prove that the angle of an n-gon is $\left(1 - \dfrac{2}{n}\right)\pi$.

   (b)   Suppose at each vertex of a tessellation there is an a_1-gon, an a_2-gon, ..., and an a_n-gon. Show that

$$\frac{1}{a_1} + \frac{1}{a_2} + \cdots + \frac{1}{a_n} = \frac{n}{2} - 1.$$

2.* Show that in addition to the regular tessellations, the following eight homogeneous tessellations exist (Figures 8.14–8.21).

   (i) $n = 3,\ a_1 = 4,\ a_2 = 8,\ a_3 = 8$ (Figure 8.14);
   (ii) $n = 3,\ a_1 = 3,\ a_2 = 12,\ a_3 = 12$ (Figure 8.15);
   (iii) $n = 3,\ a_1 = 4,\ a_2 = 6,\ a_3 = 12$ (Figure 8.16);

(iv) $n = 4, a_1 = 3, a_2 = 3, a_3 = 6, a_4 = 6$ (Figure 8.17);
(v) $n = 4, a_1 = 3, a_2 = 4, a_3 = 4, a_4 = 6$ (Figure 8.18);
(vi) $n = 5, a_1 = 3, a_2 = 3, a_3 = 3, a_4 = 4, a_5 = 4$ (Figure 8.19);
(vii) $n = 5, a_1 = 3, a_2 = 3, a_3 = 3, a_4 = 3, a_5 = 6$ (Figure 8.20);
(viii) $n = 5, a_1 = 3, a_2 = 3, a_3 = 3, a_4 = 4, a_5 = 4$ (Figure 8.21).

3.* Prove that no tessellations other than the three regular ones and those in Problem 2 exist. Note that there are seven other solutions of the equation in Exercise 1 but each of these leads to an overlapping design.

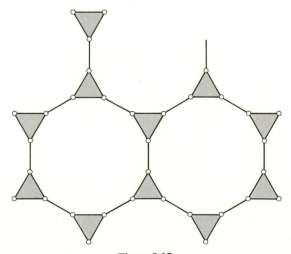

Figure 8.14

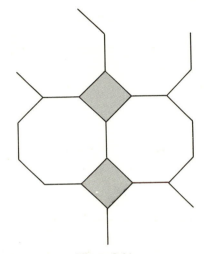

Figure 8.15

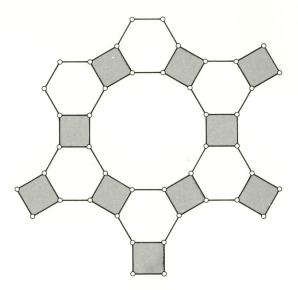

Figure 8.16

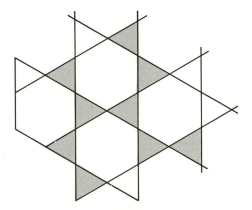

Figure 8.17

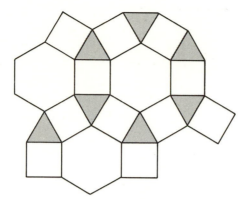

Figure 8.18

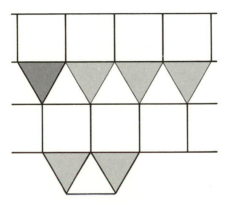

Figure 8.19

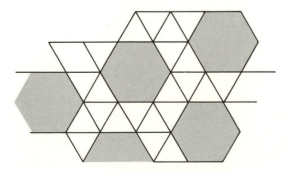

Figure 8.20

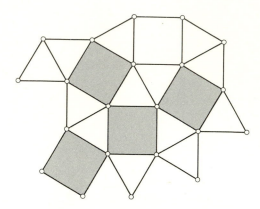

Figure 8.21

8.4 Some Equivalence Classes

Consider the set $\mathscr{R}$ of rotations of the plane about the origin. If R_θ and R_ϕ represent counterclockwise rotations of angles θ and ϕ respectively, then $R_\theta \circ R_\phi = R_{\theta+\phi}$ represents the counterclockwise rotation through the angle $\theta + \phi$. The "product" $R_\theta \circ R_\phi$ stands for the rotation R_θ followed by the rotation R_ϕ. The set $\mathscr{R}$ together with the binary operation $\circ$ is a group. This means that, for all angles θ, ϕ, and ψ,

$$(R_\theta \circ R_\phi) \circ R_\psi = R_\theta \circ (R_\phi \circ R_\psi) \qquad \text{(associative law)}$$
$$R_0 \circ R_\theta = R_\theta \circ R_0 = R_\theta \qquad \text{(existence of identity)}$$
$$R_\theta \circ R_{-\theta} = R_{-\theta} \circ R_\theta = R_0 \qquad \text{(existence of inverses).}$$

In this case the identity is a rotation through $0°$ (no rotation) and the inverse $R_{-\theta}$ of R_θ is the clockwise rotation through θ.

The ideas of this section do not require a knowledge of the algebraic properties of groups.

Consider a set S of geometrical objects in the plane. Suppose that S is partitioned into disjoint sets A, B, C,..., with $S = A \cup B \cup C \cup \cdots$ such that a subgroup of rotations leave these sets fixed. This means that if $\mathscr{R}^*$ is a subset of $\mathscr{R}$ which is also a group, and if g is an element of one of these subsets of S, then the image of g under the rotations of $\mathscr{R}^*$ is in the same subset. If these subsets A, B, C, ... are maximal subsets with these properties, they are called **equivalence classes** of S relative to the group $\mathscr{R}^*$.

The following two examples illustrate these ideas.

EXAMPLE 1   Consider a square board which may possibly contain a small circle in one or more corners. For each corner we have two choices, to place a

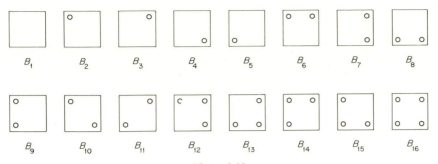

Figure 8.22

circle in the corner or leave it blank. Therefore we have 2^4 possible boards. The 16 different boards satisfying this description are shown in Figure 8.22.

If the four sides of the boards are unmarked and we cannot distinguish one side from another, then some of the above boards are equivalent under counterclockwise rotations in the plane of 90°, 180°, or 270° about the center of the square board. (These rotations from a subgroup of $\mathscr{R}$.) Two boards will be said to be equivalent if the one can be rotated in the plane about its center so as to coincide with the other. For example, B_2 rotated 270° counterclockwise about its center then coincides with B_3 so that we say B_2 and B_3 are equivalent.

We note that B_2, B_3, B_4, and B_5 are equivalent under such rotations and there are no other boards equivalent to them. We call this set of boards an "equivalence class" of boards. The reader may check that $\{B_6, B_7, B_8, B_9\}$, $\{B_{10}, B_{11}\}$, and $\{B_{12}, B_{13}, B_{14}, B_{15}\}$ are other equivalences classes. In this example we have six equivalence classes since $\{B_1\}$ and $\{B_{16}\}$ are also equivalence classes which, in these cases, consist of one element each.

EXAMPLE 2   A mark may be placed at either end or in the middle of a rod (Figure 8.23). Determine the number of equivalence classes of rods under rotations in the plane about the center of the rod.

Solution   We have 2^3 different rods since there are three locations where a mark may or may not be placed.

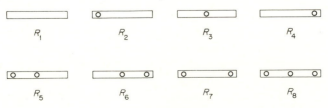

Figure 8.23

Clearly $\{R_1\}$ and $\{R_8\}$ are equivalence classes. R_2 and R_4 can be mapped into each other by a rotation of 180° in the plane about the center of the rod, as can R_5 and R_6. Also, R_3 and R_7 map into themselves under rotation. Thus there are six equivalence classes:

$$\{R_1\},\ \{R_2, R_4\},\ \{R_3\},\ \{R_5, R_6\},\ \{R_7\},\ \{R_8\}.$$

EXERCISE 8.4

1. Find the number of equivalence classes under rotations in the plane if a circle is also allowed in the center of the board of Example 1.

2. Consider an equilateral triangle which may or may not have a circle in each corner. Draw all such triangles and determine the number of equivalence classes into which they may be divided under rotations in the plane about the center of the triangles.

3. Repeat (2) if a circle is allowed in the center of the equilateral triangle.

4. Repeat (2) for (a) an isosceles triangle, (b) an ordinary triangle.

5. A rectangle (nonsquare) may or may not have a circle in each corner. Determine the number of equivalence classes under rotations in the plane about the center of the rectangle.

6. Repeat Problem 5 if a circle is allowed in the center.

9

Maps on a Sphere

We discussed ordinary plane maps in Sections 1.4, 3.3, and 5.6. In applications to cartography, a plane map represents a map on a sphere, namely the earth which is normally considered a sphere in map theory. It is clear that there is a one-to-one correspondence between plane maps that cover the whole plane (the region outside an ordinarily drawn map is considered an additional region) and maps on a sphere. To see that a map on the sphere corresponds to a map in the plane, take the plane tangent to the sphere at its "south pole" and project the map on the sphere onto the plane from the "north pole" (that is, take the intersection with the plane of all lines joining the north pole to points of the map on the sphere). This procedure may be reversed to find a map on a sphere corresponding to a map on the plane. The map obtained in this way is not the ordinary projection of the earth used in atlases. (See Exercise 9.1.)

A polyhedron is a three-dimensional geometrical object whose surface consists of plane faces that are polygons. If the plane faces are convex sets (see Section 8.1) and if the set of interior points is convex (any line segment joining two points in the set lies entirely in the set) the polyhedron is called a **convex polyhedron**. Simple examples of convex polyhedra are the cube and the tetrahedron. There is a connection between maps on a sphere and convex polyhedra. To see this, construct a sphere which contains the given polyhedron in the interior. Project (say with a light source) the vertices and edges of the polyhedron onto the surface of the sphere from a point (or light source) situated interior to the polyhedron. This yields a map on the surface of the sphere whose vertices and edges correspond to the vertices and edges of the polyhedron, and whose regions correspond to the faces of the polyhedron.

Coloring a polyhedron so that adjacent faces have different colors is thus equivalent to coloring the corresponding map on the sphere.

9.1 Euler's Formula

Euler's formula was introduced in Section 7.3 in connection with the method of averaging. It is a very important and well-known result and is a useful tool in geometry and in graph theory. Although it bears the name of Euler (eighteenth century), it was known to Descartes (seventeenth century). The formula is used in certain existence proofs (see Example 4).

Euler's formula is sometimes called Euler's polyhedron formula because it is used in connection with convex polyhedra in three-space. However in some cases the formula is valid even when a polyhedron is not convex.

EULER'S POLYHEDRON FORMULA. Let F denote the number of faces of a convex polyhedron, E the number of edges, and V the number of vertices. Then

$$F - E + V = 2.$$

A proof of this theorem is suggested in Exercise 9.1.

EXAMPLE 1   For a cube (see Figure 9.1), $F = 6$, $E = 12$, and $V = 8$. Then

$$F - E + V = 6 - 12 + 8 = 2.$$

EXAMPLE 2   For a tetrahedron, $F = 4$, $E = 6$, and $V = 4$. Then

$$F - E + V = 4 - 6 + 4 = 2.$$

Any three-dimensional convex polyhedron can be drawn in the plane, and we shall call the resulting map a **polyhedral map** or graph. Figure 9.1 shows the usual drawing of the three-dimensional cube and the corresponding polyhedral map. Note that the sixth region on the polyhedral map corresponding to the sixth face of the cube is the external, infinite region of the plane.

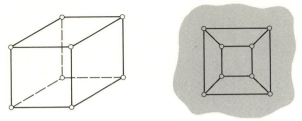

Figure 9.1

To be a polyhedral map, the map must be obtainable from some (three-dimensional) polyhedron. Not every map in the plane is polyhedral. Thus, for example, the boundary of every region of a polyhedral map must contain at least three edges since each face of the corresponding polyhedron is a convex n-gon with $n \geq 3$. In addition, every vertex of a polyhedral map must have at least three edges meeting at it since the corresponding polyhedron is three-dimensional.

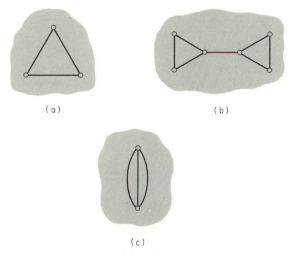

(a)   (b)

(c)

Figure 9.2

EXAMPLE 3   The graphs in Figure 9.2 are not polyhedral. In (a), each region (there are two) is bounded by three edges, but each vertex has valency two, rather than at least three. In (b), there are again vertices of valency two. In (c), each vertex is of valency three, but each of the three regions is bounded by only two edges. Thus there can be no polyhedra corresponding to these plane maps.

The constraints mentioned and illustrated above give additional constraints on V, E, and F, other than the constraint $F - E + V = 2$ of Euler's formula. If p is the average number of edges at each vertex, then $p \geq 3$, and since $pV = 2E$, we have

$$3V \leq 2E. \tag{1}$$

Similarly, if q is the average number of edges in the boundary of a region or face, then $q \geq 3$, and, since $qF = 2E$, we have

$$3F \leq 2E. \tag{2}$$

The reader should check that both (1) and (2) are violated by the map in (a) of Example 3, only (1) is violated by the map in (b), and only (2) is violated by the map in (c). It should be stressed that conditions (1) and (2) are necessary conditions for a polyhedral map but not sufficient. The map in Figure 9.3 satisfies both conditions (1) and (2) but is not polyhedral. We cannot find a polyhedron corresponding to this map.

Figure 9.3

EXAMPLE 4   Prove that there does not exist a polyhedral graph with seven edges.

Proof   Let us assume that a polyhedral map or graph with seven edges does exist, that is $E = 7$ for the map. Then, from (1), $3F \leq 2E = 14$ which implies $F \leq 4$. From (2), $3V \leq 2E = 14$ which implies $V \leq 4$. We conclude then that $F + V \leq 8$. However, from Euler's formula, $F + V = E + 2 = 9$, and so we have a contradiction. We conclude that a polyhedral graph with seven edges does not exist.

It should be pointed out that Euler's formula applies to all maps in the plane in which the edges and vertices are a connected graph and not just to polyhedral maps. In Example 5 we show that Euler's formula holds for some maps that have already been indicated as not polyhedral.

EXAMPLE 5   Check Euler's formula for the maps of Figure 9.2 and Figure 9.3.

Solution   (i)   In the map of Figure 9.2a, $F = 2$, $E = 3$, $V = 3$, and

$$F - E + V = 2 - 3 + 3 = 2.$$

(ii)   In the map of Figure 9.2b, $F = 3$, $E = 7$ $V = 6$, and

$$F - E + V = 3 - 7 + 6 = 2.$$

(iii)   In the map of Figure 9.2c, $F = 3$, $E = 3$, $V = 2$, and

$$F - E + V = 3 - 3 + 2 = 2.$$

(iv)   In the map of Figure 9.3, $F = 6$, $E = 10$, $V = 6$, and

$$F - E + V = 6 - 10 + 6 = 2.$$

In the following example, Euler's formula is applied to an enumeration problem.

EXAMPLE 6   Given a convex n-gon in the plane with all possible diagonals drawn and with no three diagonals concurrent inside the n-gon, into how many regions is the n-gon divided?

Solution   We have a map in the plane (the convex n-gon) with n outer vertices each having $n - 1$ edges meeting at it (each vertex is joined to every other vertex) and $\binom{n}{4}$ inner vertices (any selection of four outer vertices determines six edges of which exactly two cross inside the polygon). Each of the inner edges has four edges meeting at it. Thus

$$2E = \text{twice the number of edges}$$

$$= 4\binom{n}{4} + n(n - 1)$$

and

$$E = 2\binom{n}{4} + \frac{1}{2}n(n - 1).$$

If we ignore the infinite face in applying Euler's formula, the number F^* of faces or regions into which the n-gon is divided is

$$F^* = E + 1 - V$$

$$= 2\binom{n}{4} + \frac{1}{2}n(n - 1) + 1 - \binom{n}{4} - n$$

$$= \binom{n}{4} + \frac{1}{2}(n - 1)(n - 2).$$

EXAMPLE 7 Check the result of Example 6 for a convex quadrilateral, that is a convex 4-gon.

Solution If all the diagonals are drawn for a convex 4-gon, the 4-gon is divided into four regions. If we set $n = 4$ in the formula for F^* in Example 6, we obtain

$$F^* = \binom{4}{4} + \frac{1}{2}(3)(2) = 1 + 3 = 4.$$

We could, of course, have regarded this as a foregone conclusion.

EXERCISE 9.1

1. Show that there is no polyhedral map with 30 edges and 11 faces.

2. Draw two polyhedral maps with six vertices and ten edges.

3. Prove that for any polyhedral map, (i) $V \geq 2 + F/2$ and (ii) $F \geq 2 + V/2$.

4. Prove that for any polyhedral map, (i) $3F \geq E + 6$ and (ii) $3V \geq E + 6$.

5. If G is a polyhedral map with 12 vertices and 30 edges, show that *all* its faces are triangles.

6. Using the results in (4) together with $3F \leq 2E$ and $3V \leq 2E$, prove that $V \geq 4$, $F \geq 4$, and thus $E \geq 6$ for any polyhedral map.

7.* Every polyhedral map has at least two faces with the same number of edges in their boundaries.

8.* (a) Enumerate all planar maps with (i) two, (ii) three, (iii) four faces.
    (b) Consider a planar map M with V vertices, E edges, and F faces. Construct a planar map M' with $F' = F + 1$ faces, V' vertices, and E' edges by partitioning one of the faces of M. Consider all possible cases (for example, add a vertex not on an edge and join it to two vertices of M). In each of these cases verify that $V' - E' + F' = V - E + F$.
    (c) Use (a) and (b) to verify Euler's polyhedral formula for planar maps by induction.

9.   Consult an appropriate reference to find out how regions of the earth
     are represented in an atlas. In particular, find out about

   (i)   Mercator projection
  (ii)   stereographic projection
 (iii)   maps used for aerial navigation.

9.2 Regular Maps in the Plane

A map with a finite number F of faces or regions in the plane is **regular** if
each face is an n-gon for some n and if the same number m of its E edges meet
at each of its V vertices.

We use Euler's formula to investigate the existence of regular maps in the
plane. It follows from the discussion in Section 9.1 that we must solve the
system of equations

$$F - E + V = 2 \tag{1}$$

$$F = 2E/n \tag{2}$$

$$V = 2E/m \tag{3}$$

for integers m, n, F, E, and V. Substituting (2) and (3) in (1) yields

$$\frac{1}{n} + \frac{1}{m} - \frac{1}{2} = \frac{1}{E}. \tag{4}$$

The solutions of this equation which may be found by trial and error are
listed in the following Table 9.1. The values of F and V are found from (2)
and (3).

TABLE 9.1

m	n	V	E	F	Name
2	λ	λ	λ	2	Polygon (Figure 9.4)
λ	2	2	λ	λ	(Figure 9.5)
3	3	4	6	4	Tetrahedron (Figure 9.6)
3	4	8	12	6	Cube (Figure 9.7)
3	5	20	30	12	Dodecahedron (Figure 9.8)
4	3	6	12	8	Octahedron (Figure 9.9)
5	3	12	30	20	Icosahedron (Figure 9.10)

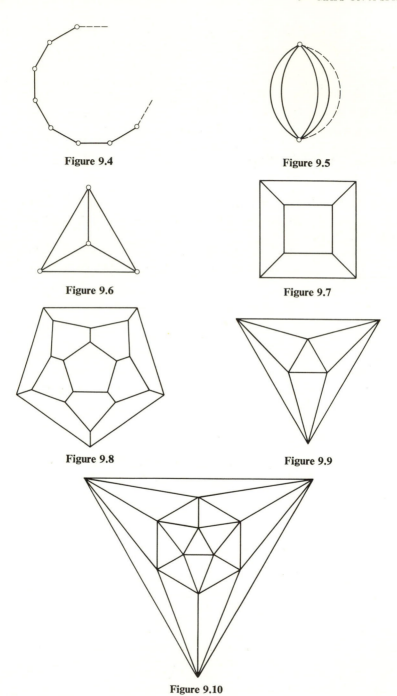

Figure 9.4

Figure 9.5

Figure 9.6

Figure 9.7

Figure 9.8

Figure 9.9

Figure 9.10

9.3 Platonic Solids

We mentioned in the introduction to this chapter that maps in the plane can be drawn as maps on a sphere and any convex polyhedron can be projected into a map on a sphere.

A **regular convex polyhedron** is a convex polyhedron whose faces are congruent polygons and which has the same number of faces meeting at each vertex. Such polyhedra are sometimes referred to as the **platonic solids**. If a regular convex polyhedron is projected onto a sphere, the corresponding map is regular. Now since all regular maps are listed in Table 9.1, there can be at most five regular convex polyhedra (platonic solids). There are exactly five and they can be easily constructed.

THEOREM   There exist exactly five regular convex polyhedra, namely, the tetrahedron, cube, dodecahedron, octahedron, and icosahedron.

EXERCISE 9.3

1.   Construct cardboard models of the five platonic solids.

Coloring Problems

We have already encountered coloring problems in our discussion of chromatic polynomials. In this chapter we again consider chromatic polynomials as well as several other coloring problems.

10.1 The Four Color Problem

Can a map on a plane or on the surface of a sphere be colored with no more than four colors so that no two adjacent countries have the same color? Each country must consist of a single connected region and adjacent countries are those having a common boundary line and not merely a single point in common.

Many excellent results in graph theory have been obtained as by-products of attempts to prove or disprove the celebrated four color conjecture which asserts that such a coloring is always possible.

The problem is not new, having attained popularity over one hundred years ago and respectability when the great British mathematician Cayley announced that he had tried and failed to solve it. Many fallacious proofs have been presented over the years and one, by Kempe in 1879, actually stood for ten years until Heawood pointed out the error in his proof in 1890. At that time Heawood proved the five color theorem,[1] showing that five colors are always sufficient to color such a map.

We will show that to prove the four color conjecture it is sufficient to establish the result for **trivalent maps**, that is, for maps with three countries at each vertex.

THEOREM 1   If every trivalent planar map can be colored in four colors, then every planar map can be colored in four colors.

Proof   The proof follows by introducing a small new country at each vertex of degree greater than three (Figure 10.1). Now the map is trivalent.

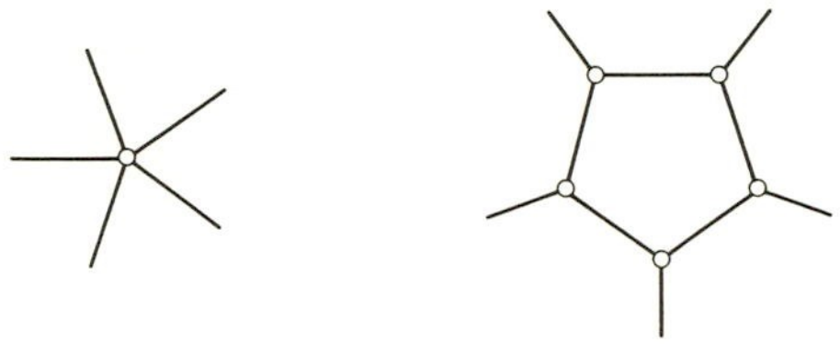

Figure 10.1

If it can be colored in four colors, we can shrink each of these new countries to points and the original map will be colored with four colors.

We will not prove the five color theorem here but rather we will prove an interesting theorem similar to ones we considered in Chapter 9 and which is used in the proof of the five color theorem.

THEOREM 2   Every planar map in which at least three edges meet at each vertex has a face with at most five edges in its boundary.

Proof   Let p be the average number of edges per vertex and q the average number of edges per face in the map. Then $p \geq 3$. If there is no such face with at most five edges, then $q \geq 6$. As in Problem 4 of Exercise 9.1 we proved $3F \geq E + 6$ so that $6F \geq 2E + 12 > 2E$. But since $q \geq 6$ and $qF = 2E$, we have $6F \leq 2E$, which is a contradiction. This proves the theorem. Note the use of the method of averaging in this proof.

While we have been mainly interested in planar graphs, we mention here the generalization of the map-coloring problem to surfaces of higher genus. A surface is of **genus** $p \geq 1$ if it can be distorted into (or is "topologically

[1] For a proof of this theorem see: F. Harary, "Graph Theory," Addison-Wesley, Reading, Massachussetts, 1969.

equivalent to") a surface obtained by adding p handles to a sphere. Thus the plane and sphere have genus 0. Figure 10.2 shows a surface of genus one; the sphere has one handle added to it. A needle, a cup, an anchor, a doughnut, and a wedding ring, all have surfaces of genus 1. A two-holed button, a coffee pot, and a two-holed pretzel have surfaces of genus two. Perhaps surprisingly, the map coloring problem has been solved for surfaces of positive genus, (the Heawood coloring theorem) while the four color conjecture seems destined to be around for some time.

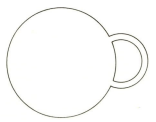

Figure 10.2

The Heawood map coloring theorem was stated as a conjecture in 1890 and proved over a period of years by Ringel and Youngs.[2] The final case was disposed of only in 1968. This theorem is stated here without proof. In the statement of the theorem the symbol $[x]$ stands for the largest integer $n \leq x$.

THEOREM 3   (Heawood map coloring theorem)   To color a map on a surface of genus p with $p \geq 1$, it can be shown that $[\frac{1}{2}(7 + \sqrt{1 + 48p})]$ colors are sufficient and may be required.

It should be pointed out that while the easy part of the four color conjecture is to show that four colors may be required, the easy part of Heawood's conjecture was to show that $[\frac{1}{2}(7 + \sqrt{1 + 48p})]$ colors are sufficient. Heawood himself did this in 1890.

Notice that if $p = 0$ in Theorem 3, the surface involved is a plane or sphere and $[\frac{1}{2}(7 + \sqrt{1 + 48(0)})] = 4$ so that $p = 0$ in the Heawood theorem is equivalent to the four color theorem.

A number of results regarding the four color conjecture have been announced recently but it has not been resolved. As this book goes to press, the best result known seems to be that all maps with fewer than 36 countries can be colored with four colors.

---

[2] G. Ringel and J. W. T. Youngs, Solution of the Heawood Map-Coloring Problem, *Proceedings National Academy of Science, U.S.A.*, **60** (1968), 438–445.

1. How many colors may be needed to color a map on a torus ($p = 1$) or a
pretzel ($p = 2$)?

2.* What is the least number of colors required to color the faces of each of
the regular polyhedral graphs of Section 9.2? (Note that the faces of the
dodecahedron cannot be colored with three colors even though there is no
set of four faces all touching each other.)

3.* Show that any polyhedral map with fewer than twelve faces must have a
face with at most four edges in its boundary.

10.2 Coloring Graphs

In Section 1.4 we pointed out that map-coloring problems are special
cases of graph-coloring problems. The graph G corresponding to a map can
be obtained by taking a point interior to each region of the map as a vertex
of G and by joining two vertices of G by an edge if the corresponding regions
of the map have a boundary in common. The condition for a map coloring,
namely that two regions with a common boundary must be given different
colors, becomes the condition that two adjacent vertices of the graph G
must be colored differently, that is, the condition for a proper coloring of the
graph G.

If a map (representing countries each consisting of a single connected
region with adjacent countries having a boundary line and not merely a
point in common) is drawn in the plane, then the corresponding graph G is a
planar graph, that is, a graph which can be drawn in the plane in such a way
that edges intersect only at the vertices. It was pointed out in Section 1.3 that
the graph in Figure 1.21 is not a planar graph.

We see therefore that map-coloring problems are equivalent to problems
of coloring planar graphs. Recalling that $P(G, \lambda)$ stands for the number of
ways of coloring graph G with λ or fewer colors, the famous four color
conjecture, that any planar map can be colored in at most four colors, can
be reformulated as

$$\text{``} P(G, 4) \neq 0 \text{ if } G \text{ is planar,''}$$

since $P(G, 4) = 0$ means that G cannot be colored in four or fewer colors.

1. Draw the graphs corresponding to the maps associated with the tetrahedron and the cube.

2.   Determine the class of maps which correspond to trees.

3.   Determine the map corresponding to the *n*-gon considered as a graph.

4.   A wheel is a graph obtained by adjoining to an *n*-gon graph one vertex and the edges obtained by joining this vertex to the vertices of the *n*-gon. Determine the planar map corresponding to such a graph.

10.3 More about Chromatic Polynomials

Various short cuts are available whereby the calculation of chromatic polynomials (previously studied in Sections 1.4, 3.3, and 5.6) can be facilitated. Some of these are contained in the following theorems. These ideas illustrate the way in which any mathematical technique can be developed into an effective tool by further study.

THEOREM 1   If a graph *G* consists of *k* components $G_1, G_2, \ldots, G_k$, then

$$P(G, \lambda) = P(G_1, \lambda)P(G_2, \lambda) \cdots P(G_k, \lambda).$$

(This was stated for $k = 2$ in Theorem 4 of Section 5.6)

Proof   Since the components are disjoint, the coloring of each is quite independent of the coloring of the others. Hence the number of ways of coloring the whole graph is simply the product of the numbers of colorings of the separate components (multiplication principle).

In the above theorem we used the term "subgraph" and assumed that the reader had an intuitive idea of the meaning of the term. A graph *H* is a subgraph of a graph *G* if the vertices of *H* are vertices of *G* and if the edges of *H* are edges of *G*. In Figure 10.3, H_1 is a subgraph of *G* but H_2 is not.

In the following theorem we use the term "overlap in a complete graph" as applied to two subgraphs *X* and *Y* of a graph *G*. We say that two subgraphs *X* and *Y* of a graph *G* **overlap in a complete graph** if the sets of vertices of *X* and

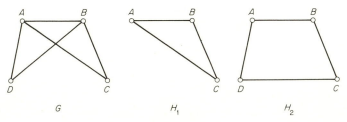

Figure 10.3

Y have k vertices in common and if every pair of these k vertices is joined by an edge in both X and Y.

EXAMPLE 1   The subgraphs X and Y of the graph G (Figure 10.4) overlap in the complete graph on three vertices.

Figure 10.4

THEOREM 2   If two graphs X and Y overlap in a complete graph on k vertices, then the chromatic polynomial of the graph G which is the collection of all vertices and edges of X and Y is

$$\frac{P(X, \lambda)P(Y, \lambda)}{\lambda^{(k)}}.$$

Proof   Since the common part of X and Y is the complete graph on k vertices, the number of ways of coloring it will be $\lambda^{(k)}$ (Theorem 2, Section 5.6). For every coloring of these k vertices there will be $P(X, \lambda)/\lambda^{(k)}$ ways of coloring the remaining vertices of X and $P(Y, \lambda)/\lambda^{(k)}$ ways of coloring the remaining vertices of Y. Hence the total number of colorings is

$$\lambda^{(k)} \cdot \frac{P(X, \lambda)}{\lambda^{(k)}} \cdot \frac{P(Y, \lambda)}{\lambda^{(k)}} = \frac{P(X, \lambda) \cdot P(Y, \lambda)}{\lambda^{(k)}}$$

as required.

EXAMPLE 2   The graph X of Example 1 contains a complete subgraph on three vertices. Show that the number of ways of coloring the two remaining vertices is $P(X, \lambda)/\lambda^{(3)}$ (as in the proof of Theorem 2).

Solution   Using the iterative technique of Section 5.6 we may write (Figure 10.5).

Figure 10.5

Thus

$$P(X, \lambda) = \lambda^{(5)} + 3\lambda^{(4)} + 2\lambda^{(3)}$$
$$= \lambda^{(3)}(\lambda^2 - 4\lambda + 5)$$

so that

$$P(X, \lambda)/\lambda^{(3)} = \lambda^2 - 4\lambda + 5.$$

Now consider the colorings of vertices A and B in X once vertices C, D, and E have been colored (Figure 10.5). There are three possibilities (i) C, D, and A are colored differently; (ii) A and D are the same color, C different, (iii) A and C are the same, D different. In the first there are $\lambda - 3$ choices for A and $\lambda - 3$ choices for B; in the second, one choice for A, $\lambda - 2$ for B, and in the third, one choice for A and $\lambda - 2$ for B. The total number of colorings of A and B is then

$$(\lambda - 3)^2 + (\lambda - 2) + (\lambda - 2) = \lambda^2 - 4\lambda + 5,$$

as asserted.

We next apply Theorem 2 to prove the following theorem which was stated as problem (4) in Exercise 1.4. We defined a **tree** in Section 1.3.

THEOREM 3   The chromatic polynomial of a tree having n vertices is $\lambda(\lambda - 1)^{n-1}$.

Proof   We can build up any tree by starting with a single edge and adding edges one by one, each added edge having one vertex in common with the tree so far constructed. An example is shown in Figure 10.6. Hence by Theorem 2

Figure 10.6

with $k = 1$, the chromatic polynomial of the new tree (having the added edge) is obtained from that of the old tree by multiplying the latter polynomial by the chromatic polynomial of the edge, $\lambda(\lambda - 1)$, and dividing by $\lambda^{(1)}$ or λ. This means that the addition of each edge multiplies the previous chromatic polynomial by $\lambda - 1$. Since we started with a single edge [chromatic polynomial $\lambda(\lambda - 1)$], the theorem follows by induction.

THEOREM 4   The chromatic polynomial of an n-gon is

$$(\lambda - 1)^n + (-1)^n(\lambda - 1).$$

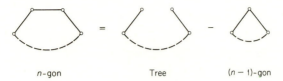

n-gon	Tree	(*n* − 1)-gon

Figure 10.7

Proof   Let $P_n(\lambda)$ denote the chromatic polynomial of an *n*-gon. Then (Figure 10.7) we may express the chromatic polynomial $P_n(\lambda)$ in terms of the chromatic polynomial of a tree and the chromatic polynomial of an (*n* − 1)-gon, using the technique of Section 5.6. Hence we have

$$P_n(\lambda) = \lambda(\lambda - 1)^{n-1} - P_{n-1}(\lambda)$$

by Theorem 3. This can be written

$$P_n(\lambda) - (\lambda - 1)^n = (\lambda - 1)^{n-1} - P_{n-1}(\lambda)$$

from which it follows that $(-1)^n\{P_n(\lambda) - (\lambda - 1)^n\}$ is a constant which can be found by setting *n* = 3. Then

$$-\{P_3(\lambda) - (\lambda - 1)^3\} = -\{\lambda(\lambda - 1)(\lambda - 2) - (\lambda - 1)^3\}$$
$$= \lambda - 1,$$

since $P_3(\lambda) = \lambda(\lambda - 1)(\lambda - 2)$. Then

$$P_n(\lambda) - (\lambda - 1)^n = (-1)^n(\lambda - 1)$$

or

$$P_n(\lambda) = (\lambda - 1)^n + (-1)^n(\lambda - 1),$$

as required. Note that we can obtain the same result more simply if we regard a line segment (2-gon) as a degenerate *n*-gon and substitute *n* = 2 instead of *n* = 3 in the above.

Properties of Chromatic Polynomials[3]

We shall now list without proof some properties of the chromatic polynomial $P(G, \lambda)$ of a graph *G*. We let *n* denote the number of vertices of *G*.

Theorem 5   The degree of $P(G, \lambda)$ is *n*.

Theorem 6   The coefficient of λ^n in $P(G, \lambda)$ is 1.

---

[3] For a further discussion, see: R. C. Read, An Introduction to Chromatic Polynomials, *Journal of Combinatorial Theory*, **4** (1968), 52–71.

THEOREM 7 $P(G, \lambda)$ has no constant term.

THEOREM 8   The terms in $P(G, \lambda)$ alternate in sign.

THEOREM 9   The absolute value of the second coefficient of $P(G, \lambda)$ is the number of edges in G.

THEOREM 10   If G is a connected graph, then $P(G, \lambda) \leq \lambda(\lambda - 1)^{n-1}$ for any positive integer λ.

THEOREM 11   A necessary and sufficient condition for a graph G having n vertices to be a tree is that $P(G, \lambda) = \lambda(\lambda - 1)^{n-1}$.

THEOREM 12   If G is connected then the absolute value of the coefficient of λ^r in $P(G, \lambda)$ is not less than $\binom{n-1}{r-1}$.

COROLLARY   The smallest number r such that λ^r has a nonzero coefficient in $P(G, \lambda)$ is the number of components of G.

UNSOLVED PROBLEMS

There are many unsolved problems concerning chromatic polynomials which can be easily formulated; we shall mention only a few. First and foremost is the question, "What makes a polynomial chromatic"? We have mentioned various necessary conditions for a polynomial to be the chromatic polynomial of some graph (Theorems 5, 6, 7, 8, 10), but none of them is sufficient. For example, the polynomial

$$\lambda^4 - 3\lambda^3 + 3\lambda^2$$

satisfies these conditions but is not the chromatic polynomial of any graph. The problem of characterizing chromatic polynomials is unsolved. Another unsolved problem in a similar vein is that of determining what numbers can be roots of chromatic polynomials.

A property that is very noticeable when one examines a few chromatic polynomials is that the coefficients first increase in absolute magnitude, and then decrease; two successive coefficients may be equal, but it seems that one never observes a coefficient flanked by larger coefficients, and it is natural to conjecture that the coefficients always behave in this way. It is fairly easy to show that the coefficients are bounded in absolute magnitude by the corresponding coefficients in the chromatic polynomial of the complete graph on the same number of vertices (the proof of this will be left as an exercise for the reader); and certainly these upper bounds first increase and then decrease. But whether this is true for all chromatic polynomials is still an open question.

Again, this increase and decrease in the coefficients suggests that for large values of *n* the coefficients in the chromatic polynomials of *most* graphs on *n* vertices might approximate some well-known unimodal statistical distribution.

It is clear that distinct graphs may have the same chromatic polynomial. For example, all trees with *n* vertices have the same chromatic polynomial. Less trivially, the distinct graphs in Figure 10.8 have the same chromatic polynomial.

Figure 10.8

This prompts the question "What is a necessary and sufficient condition for two graphs to have the same chromatic polynomial"? This question is also unsolved.

EXERCISE 10.3

1. A square with one diagonal can be decomposed into two triangles which overlap in a complete graph on two vertices. Use this observation and Theorem 2 to find its chromatic polynomial.

2. Prove Theorem 5.

3. Prove Theorem 6.

4. Prove Theorem 7.

5. Prove Theorem 11.

6.* A wheel is the product of a vertex and an *n*-gon (see Exercise 4 in Section 10.2). Find the chromatic polynomial of a wheel.

10.4 Chromatic Triangles

Instead of coloring the vertices of a graph we are often interested in coloring the edges. (Even the four color problem can be reformulated as an edge-coloring problem.) In this section we consider an interesting idea sometimes referred to as the **problem of Ramsey**. A triangle is **chromatic** if each of its edges has the same color.

THEOREM 1   Let five points in the plane, no three collinear, be joined by ten line segments which form ten triangles. There exists a coloring of the edges in red and blue so that no triangle is chromatic.

Proof   Color a pentagon through the points red, and the remaining line segments blue, for example, to get a required coloring.

However, if six points are used instead of five, the situation is different.

THEOREM 2   Let six points in the plane, no three collinear, be joined by fifteen line segments which form twenty triangles. If the line segments are colored either red or blue in any way, there will always be a chromatic triangle.

Proof   This is easily proved by considering the five line segments through any point *P*. Three of these must be colored the same, say red, by the Dirichlet drawer principle. Then the triangle determined by the end points of these three red segments must be blue (chromatic) or if one of its sides is red, it determines a chromatic triangle with two of the three red lines through *P*.

EXERCISE 10.4

1.   Prove that six is the smallest number of points in the plane, no three collinear, having the chromatic triangle property.

2.   If 17 points in a plane, no three collinear, are joined by red, white, or blue line segments, prove that a chromatic triangle results.

3.   Show that among any six persons it is always possible to find either three who are mutually acquainted or three complete strangers.

4.*   Given seven points joined in pairs by red or blue lines, show that there are at least three chromatic triangles.

5.*   If all the diagonals of a 66-gon are drawn and all edges colored either red, blue, green, or gold, prove that there will result a chromatic triangle.

6.* Chromatically speaking, what is the next number in the sequence
{6, 17, 66, ...}?

10.5 Sperner's Lemma

The work of this section is akin to the problem of coloring graphs. As
a preliminary to Sperner's lemma we prove the following theorem.

THEOREM 1   Let points interior to the line segment AB be labeled A or B
in any way (Figure 10.9). Then there is an odd number of smaller line segments labeled AB or BA.

Figure 10.9

Proof An inductive proof can be given. Add one point to AB. If this
point is labeled either A or B, exactly one of the resultant smaller segments is
labeled AB. Assume the theorem is true if n points are taken in the segment
AB and suppose the number of segments labeled AB or BA is an odd number
v_n. Select an additional point X. If X is taken in a segment AA then $v_{n+1} =$
v_n if X is labeled A, and $v_{n+1} = v_n + 2$ if X is labeled B. In both cases v_{n+1} is odd.
(Similarly this is true if X is taken in a segment BB.) If X is taken in a segment
AB or BA, then $v_{n+1} = v_n$ for either labeling of X. The details of the induction
can be added.

A combinatorial type proof can also be given.

As above, select n points in the given segment. Suppose k of these points
are labeled A (and $n - k$ are labeled B). There are $n + 1$ segments, γ_n of which
are labeled AB or BA. Count the end points of the smaller segments which are
labeled A in two ways: there will be $2k + 1$ since there are k interior points
labeled A, and there will be $\gamma_n + 2a$ where a is the number of segments
labeled AA. Thus $\gamma_n = 2(k - a) + 1$ which is odd.

We are now in a position to prove the following result.

SPERNER'S LEMMA (1928).   Consider a triangle ABC which has been
partitioned into smaller triangles by selecting any number of points on the
sides of the triangle or in its interior as in Figure 10.10. Label the points on
the side AB either A or B, label points on the side BC either B or C, label
points on the side CA either C or A, and label the interior points A, B, C in
any way. Then the number of small triangles labeled ABC is odd.

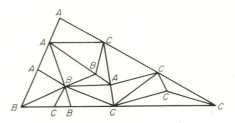

Figure 10.10

Proof of Sperner's Lemma   Consider all the small triangles and count all the edges labeled *AB*. If an interior segment is labeled *AB*, then it is counted twice; an exterior segment *AB* is counted only once. If a small triangle is not labeled *ABC* then either none of its sides is labeled *AB* or exactly two sides are labeled *AB*. Thus if *T* is the number of small triangles labeled *ABC*, *T'* the number of small triangles labeled *AAB* or *ABB*, and if γ_e is the number of exterior segments labeled *AB* and γ_i the number of interior segments labeled *AB*, then

$$T + 2T' = \gamma_e + 2\gamma_i$$

so that $T = \gamma_e + 2(\gamma_i - T')$. But by Theorem 1, γ_e is odd implying that *T* is odd.

COROLLARY   There exists at least one small triangle labeled *ABC*.

Kenneth Berman proved Theorems 2 and 3 below while a Grade 11 high school student. These theorems are similar to Sperner's lemma but no labeling is involved.

Place points along the perimeter and in the interior of a polygon. Join these points to form triangles as you will. This produces a triangulation of the polygon. In a triangulation the triangles connect only at vertices. Such triangulations are called **simplicial**. For example, consider the quadrilateral *ABCD* as shown in Figure 10.11. In this case, point *E* is on the side *AC* of triangle *ABC*. The quadrilateral *ABCD* has not been triangulated. Examples of acceptable triangulations are given in Figure 10.12, 10.13, and 10.14.

In Figure 10.12 there are nine triangles within the large triangle and five line segments along the perimeter and both of these numbers are odd. In Figure 10.13 there are twelve triangles within the given hexagon and six line segments along the perimeter and both of these numbers are even. There are nine triangles in the polygon in Figure 10.14 and eleven line segments along its perimeter, again both odd numbers involved. The results of these examples are generalized in Theorem 2. The term "parity" is used in the statement of the theorem. If two numbers are both even or both odd we say that they have the same **parity**.

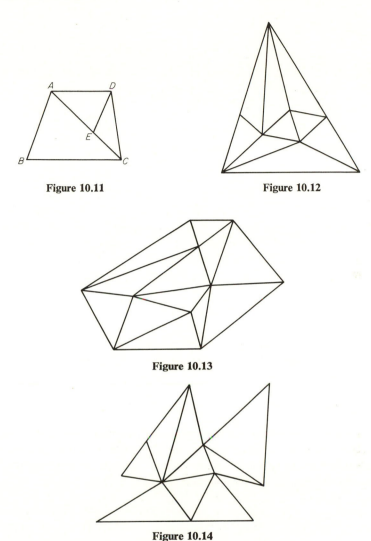

Figure 10.11

Figure 10.12

Figure 10.13

Figure 10.14

THEOREM 2 Consider any simplicial triangulation of a polygon P. Let N_P be the number of line segments on the perimeter of P and N_T the number of triangles in the triangulation of P. Then N_P and N_T have the same parity.

Proof Let N_I be the number of interior line segments. Since each line segment on the perimeter belongs to one triangle while each interior line segment is on the side of two triangles, the total number of sides of all the triangles in the triangulation is both $3N_T$ and $2N_I + N_P$. Thus

$$2N_I + N_P = 3N_T.$$

Therefore

$$N_P - N_T = 2N_T - 2N_I = 2(N_T - N_I),$$

that is, $N_P - N_T$ is an even number. However, the difference between two integers is even only if the integers are both odd or both even. Thus the theorem is proved.

Theorem 2 involved dividing a polygon into triangles. A similar more general result will hold if we divide the polygon into polygons as in Figure 10.15.

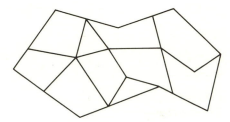

Figure 10.15

THEOREM 3   Let a polygon P be divided into polygons. Then N_P, the number of line segments on the perimeter, and N_O the number of interior polygons with an odd number of sides, have the same parity.

The reader is asked to prove this theorem in Problem 6, Exercise 10.5.

COROLLARY   Suppose all the diagonals of a convex polygon with n sides are drawn to decompose the polygon. Then N_O, the number of subpolygons produced which have an odd number of sides, has the same parity as n. This is illustrated in Figure 10.16 for the case $n = 6$.

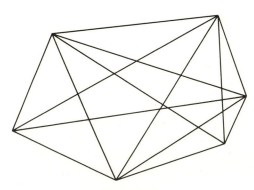

Figure 10.16

EXERCISE 10.5

1. Show that if the end points of a line segment are both labeled AA and interior points are labeled A or B in any way, then either there are no segments labeled AB or BA, or the number of segments labeled AB or BA is even.

2. Suppose the vertices of the triangles of a triangulated triangle are labeled A, B, C in any way (exterior vertices also). Show that (i) if the number of exterior segments labeled AB is odd then there is an odd number of small triangles labeled ABC; (ii) if the number of exterior segments labeled AB is even then either there is no small triangle labeled ABC or there is an even number of triangles labeled ABC.

3. If the exterior triangle in Problem (2) is labeled AAA, what can be said about the parity of the number of triangles labeled ABC?

4.* Points are taken on the circumference of a circle and labeled A, B, or C in any way. Let λ, μ, υ denote the number of arcs labeled AB (or BA), BC (or CB), and CA (or AC), respectively. Prove that λ, μ, υ are all even or all odd.

5.* *Alternative proof of Sperner's lemma*   Label the edges of the small triangles 0 or 1 depending on whether the end points are labeled the same or differently. Then

   (i)   The sum of the edge labels is 3 if a triangle is labeled ABC and is 0 or 2 otherwise.
   (ii) The sum of the labels on all triangles is odd.
   (iii) Sperner's lemma follows from (i) and (ii).

6.* Prove Theorem 3 of this section.

Finite Structures

There are finite systems studied in other branches of mathematics which are useful in combinatorics. A few such systems are described in this chapter. The discussions involve many existence arguments.

11.1 Finite Fields

The idea of a finite field was introduced in Section 7.1. It was shown in Example 4 of that section that there exists a finite field with two elements. In this section we discuss finite fields further and give several constructions for finite fields. The reader is familiar with at least three examples of infinite fields: the sets of rational numbers, real numbers, and complex numbers, under the operations of ordinary addition and multiplication.

A field $\mathscr{F}$ consists of a set F of elements closed under two binary operations, addition and multiplication, defined on these elements. We sometimes refer to $\mathscr{F}$ as a triple $(F, +, \cdot)$ where F is the set of elements and $+, \cdot$ are the symbols indicating the binary operations of addition and multiplication, respectively. The operations of addition and multiplication are called binary operations because each acts on a given pair of elements to produce a third element (which may be the same as either of the given elements). The set F is closed under a binary operation if the element associated with a pair of elements of F under this operation is an element of F. Thus, in a field $\mathscr{F}$, we associate with each pair of elements a and b of F, unique elements $a + b$ and $a \cdot b$ (or ab) called the sum and product respectively of the two elements. The operations of addition and multiplication must satisfy the following axioms or laws or rules for every a, b, and c in F.

ASSOCIATIVE LAWS

$$a + (b + c) = (a + b) + c; \qquad a(bc) = (ab)c.$$

COMMUTATIVE LAWS

$$a + b = b + a; \qquad ab = ba.$$

EXISTENCE OF IDENTITIES   Special elements 0 and 1 exist in F such that

$$a + 0 = a; \qquad a1 = a.$$

EXISTENCE OF INVERSES   Elements $-a$ and a^{-1} (if $a \neq 0$) exist in F such that

$$a + (-a) = 0; \qquad aa^{-1} = 1.$$

DISTRIBUTIVE LAW

$$a(b + c) = ab + ac.$$

The field $\mathscr{F}$ is **finite** if the cardinality of F is finite.

The rules listed above are usually taken for granted in ordinary arithmetic, that is, in the infinite fields of rational and real numbers. However, it cannot be assumed from the axioms that there exists a finite field. Axioms do not prove existence. Of course we know that at least one finite field exists once we have verified that the system defined in Example 4 of Section 7.1 does, in fact, satisfy all the above axioms.

We shall now give a number of constructions for finite fields. One way of constructing a field is by specifying the elements and the addition and multiplication tables for the elements. We illustrate this method in Example 1.

EXAMPLE 1   Let $F = \{0, 1, 2\}$ and let addition and multiplication be defined as in Table 11.1. It is left as an exercise for the reader to verify that all the rules for a field hold in this system (Problem 1 of Exercise 11.1).

TABLE 11.1

+	0	1	2	·	0	1	2
0	0	1	2	0	0	0	0
1	1	2	0	1	0	1	2
2	2	0	1	2	0	2	1

Any symbols, for example a, b, and c, could have been used in Example 1 instead of 0, 1, and 2. However, the more familiar symbols 0, 1, and 2 serve as a mnemonic for remembering the tables. This statement will become clearer with later examples.

The reader may wonder how the addition and multiplication tables were obtained in Example 1. If it were by trial and error, surely this would be an unsatisfactory way of creating a field. A simple and fairly general method of constructing a finite field involves addition and multiplication of integers, *modulo an integer m*. The expression $a + b \equiv c$ (mod m) where $0 \leq c < m$, means that c is the least nonnegative remainder when the sum $a + b$ is divided by m. Similarly, $ab \equiv d$ (mod m) with $0 \leq d < m$ means that d is the least nonnegative remainder when the product ab is divided by m. Thus, for example, the addition and multiplication in Example 1 are addition and multiplication modulo 3. The proof of the following theorem is left as an exercise for the reader (Problem 2 of Exercise 11.1).

THEOREM 1   Let p be a prime integer, let $F_p = \{0, 1, \ldots, p - 1\}$, and let $+$ and $\cdot$ respectively denote addition and multiplication modulo p of the elements of F_p. Then $\mathscr{F}_p = (F_p, +, \cdot)$ is a finite field.

The field constructed in Example 1 is thus $\mathscr{F}_3$. The construction of Theorem 1 produces an infinite number of finite fields since the number of primes is infinite. It is left as an exercise for the reader to prove that the construction does not produce a finite field if p is not a prime. This construction does not produce all finite fields. In the following example, a finite field containing nine elements is constructed using a method which is often used for constructing complex numbers from real numbers.

EXAMPLE 2   Let $\mathscr{F}_3$ denote the field constructed in Example 1 (by Theorem 1 with $p = 3$), and let i denote an abstract symbol with the property that $i^2 = -1$, that is, $i^2 = 2$ in $\mathscr{F}_3$. Let F denote the set of nine elements of the form $x + iy$, with $x, y \in \mathscr{F}_3$, and let addition and multiplication be defined by ordinary addition and multiplication of the algebraic expressions and the replacement of i^2 by 2. These are expressed in symbols by the following identities.

$$(x_1 + iy_1) + (x_2 + iy_2) = (x_1 + x_2) + i(y_1 + y_2),$$
$$(x_1 + iy_1)(x_2 + iy_2) = (x_1 x_2 + 2y_1 y_2) + i(x_1 y_2 + x_2 y_1).$$

Two elements of F are equal if they are identical, that is,

$$x_1 + iy_1 = x_2 + iy_2 \qquad \text{if and only if} \qquad x_1 = x_2 \quad \text{and} \quad y_1 = y_2.$$

It is left to the reader to verify that $(F, +, \cdot)$ is a finite field containing nine elements.

Some objection can be raised to the construction given in Example 2 since we seem to be pulling the symbol i out of a hat. We are describing i rather than showing its existence. Of course, if one accepts the analogous construction for complex numbers, one must also accept the present construction.

The reader will also have reservations if he tries to carry out the same construction starting with $\mathscr{F}_2$ rather than $\mathscr{F}_3$ (Problem 6, Exercise 11.1) particularly when he finds that the construction does not yield a field in this case. This then raises another objection to the construction; it can not easily be generalized. However the construction can be modified and in the next example we show how to construct a finite field containing p^2 elements for any prime p.

By an $\mathscr{F}_p$ **quadratic** we mean an expression of the form

$$Q(x) = ax^2 + bx + c$$

where a, b, and c are elements of F_p and x is a formal mark. The quadratic $Q(x)$ is called **irreducible** over the field $\mathscr{F}_p$ if $Q(t) \neq 0$ for any element t in F_p.

THEOREM 2   Let $Q(x)$ denote an irreducible $\mathscr{F}_p$-quadratic. Let G_p denote the set of linear polynomials over $\mathscr{F}_p$, that is, expressions of the form $ax + b$ where $a, b \in F_p$ and x is a formal mark. Define addition and multiplication of elements of G_p as follows:

$$(a_1 x + b_1) + (a_2 x + b_2) = (a_1 + a_2)x + (b_1 + b_2),$$
$$(a_1 x + b_1)(a_2 x + b_2) = c_1 x + d_1 = c_2 x + d_2$$

where $c_1 x + d_1$ is the remainder when the ordinary product (quadratic) on the left side is divided by $Q(x)$ and c_2 and d_2 are the least nonnegative integers congruent to c_1 and d_1 respectively, modulo p. Then $(G_p, +, \cdot)$ is a finite field containing p^2 elements.

The proof that all of the axioms for a field hold in $(G_p, +, \cdot)$ may be checked by the reader. The existence of an irreducible quadratic over any field $\mathscr{F}_p$ is also left as an exercise for the reader (Problem 9 of Exercise 11.1).

EXAMPLE 3   The quadratic $Q(x) = x^2 + 1$ is irreducible over the field $\mathscr{F}_3$ since $Q(0) = 1$, $Q(1) = Q(2) = 2$. The set G_3 is given by

$$G_3 = \{0, 1, 2, x, x + 1, x + 2, 2x, 2x + 1, 2x + 2\}.$$

We may construct addition and multiplication tables for these nine elements according to the definitions in Theorem 2. For example,

$$(x + 2) + (2x + 1) = 3x + 3 = 0$$
$$(x + 1) + (x + 2) = 2x + 3 = 2x$$
$$(2x + 1)(2x + 2) = 4x^2 + 6x + 2$$
$$= (x^2 + 1)(4) + 6x - 2$$
$$= 6x - 2 \qquad \text{(by definition)}$$
$$= 1 \qquad \text{(by definition)}$$

$$x(2x + 1) = 2x^2 + x$$
$$= (x^2 + 1)2 + (x - 2)$$
$$= x - 2 \qquad \text{(by definition)}$$
$$= x + 1 \qquad \text{(by definition)}.$$

The reader should construct addition and multiplication tables in this example and verify that $(G_3, +, \cdot)$ satisfies all the axioms for a field.

If we start with a polynomial that is not irreducible, the construction does not yield a field.

It is now easy to explain why the method of Example 2 works in the case of a field with three elements, but does not work in the case of two elements (Problem 6, Exercise 11.1). It is because the quadratic $Q(x) = x^2 + 1$ is an irreducible quadratic over the field $\mathscr{F}_3$, as we checked in Example 3, but not over $\mathscr{F}_2$ since $Q(1) = 1 + 1 = 0$ in $\mathscr{F}_2$.

The construction in Theorem 2 can be generalized to obtain a finite field called the Galois field $GF(p^n)$ containing p^n elements for any positive integer n. It can be shown that there are no other finite fields, that is, every finite field is isomorphic or equivalent to one of these fields $GF(p^n)$.

It is easy to prove that every finite field has p^n elements for some prime p.

THEOREM 3   If k is the cardinality of a finite field $\mathscr{F} = (F, +, \cdot)$, then $k = p^n$ for some prime p and some nonnegative integer n.

Proof   Consider the sequence 1, $2 = 1 + 1$, $3 = 1 + 1 + 1$, ... Since this sequence must be finite in F, there must be some first repeated number, that is, $m = n$ with $m > n$, and $p = m - n = 0$. This integer p must be a prime for otherwise $p = rs = 0$ for some integers r and s and this is not possible in a field (a field has no divisors of zero). Using the rules for addition and multiplication in $\mathscr{F}$, it can be shown that the elements of F corresponding to the symbols $0, 1, 2, \ldots, p - 1$, add and multiply like integers modulo p. Thus it follows that $\mathscr{F}_p$ is a subfield of $\mathscr{F}$. If there is an additional element α_1 in $\mathscr{F}$, there are p^2 different elements in $\mathscr{F}$ of the form $\alpha_1 x_1 + x_2$ with $x_1, x_2 \in \mathscr{F}_p$. If there is yet another additional element α_2 in $\mathscr{F}$, there will be p^3 different elements of the form $\alpha_1 x_1 + \alpha_2 x_2 + x_3$ in $\mathscr{F}$, with $x_1, x_2, x_3 \in \mathscr{F}_p$. Continue in this way until all the elements of F have been expressed in the form

$$\alpha_1 x_1 + \alpha_2 x_2 + \alpha_3 x_3 + \cdots + \alpha_{n-1} x_{n-1} + x_n$$

for some positive integer n, with all $x_i \in \mathscr{F}_p$. There will be p^n such elements.

EXERCISE 11.1

1. Verify that the axioms for a field are satisfied by the system defined in Example 1.

2. Prove Theorem 1.

3. Prove that $(F_m, +, \cdot)$ is not a finite field if p is replaced by a nonprime integer m in Example 2. (Such an algebraic system is called a **ring**.)

4. Verify that the field constructed in Example 1 is isomorphic, that is, equivalent to $\mathscr{F}_3$.

5. Verify that the system of nine elements constructed in Example 2 is a finite field.

6. Try to construct a field containing four elements by the method of Example 2. Show that the system obtained is not a field by showing that one of the rules does not hold.

7. Use the method of Example 2 to construct a field with 49 elements.

8. Construct the addition and multiplication tables in Example 3.

9.* (a) Prove that if the F_p-quadratic $ax^2 + bx + c$ has zeros x_1, x_2, then it can be factored, as in the case of real quadratics, into the form

$$ax^2 + bx + c = a(x - x_1)(x - x_2).$$

   (b) Deduce the existence of an irreducible quadratic (by the method of exhaustion).

10.* (a)   Prove Theorem 2.

      (b)   State the generalization of Theorem 2 (employing irreducible polynomials of degree n).

11.* (a)   Show that $x^2 + x + 1$ is an irreducible quadratic over $GF(2)$.

      (b)   Use (a) to construct $GF(4)$ using Theorem 2.

12.* Verify that the p^n elements constructed in the proof of Theorem 3 are, in fact distinct.

11.2 The Fano Plane

For many centuries Euclidean geometry, which is the geometry normally studied in high school mathematics, was considered the only possible geometry. Then non-Euclidean geometries were introduced in the eighteenth century but all these geometries were *infinite* geometries. It was not until the end of the nineteenth century that Fano showed that it was possible to have a geometry with a finite number of points and lines. In this section, we explain the concept of a projective geometry and in particular the concept of a finite projective plane. We begin with a practical problem.

EXAMPLE   There are seven firemen assigned to a particular station. Three firemen are required for night duty. Construct a fair weekly schedule.

Solution   A fair schedule implies that each man works the same number of nights. Since there are 3×7 man-nights required, each fireman must work three nights. Is such a schedule possible? The answer, which is in the affirmative, may be found by trial and error. A possible schedule is given in Table 11.2. For convenience, the firemen are labeled $F_1, F_2, \ldots, F_7$.

TABLE 11.2

Night Schedule	
Sunday	F_1 F_2 F_4
Monday	F_2 F_3 F_5
Tuesday	F_3 F_4 F_6
Wednesday	F_4 F_5 F_7
Thursday	F_5 F_6 F_1
Friday	F_6 F_7 F_2
Saturday	F_7 F_1 F_3

It should be pointed out that merely stating a problem does not ensure the existence of a solution. For example, notice that the schedule proposed in Table 11.2 has the property that every pair of firemen serves together on exactly one night. This property cannot be preserved in a schedule requiring four firemen out of seven to work each night.

The schedule in Table 11.2 is represented diagrammatically in Figure 11.1. Each of the firemen is represented by a point and the nights of the schedule are represented by the lines and diagonals of a triangle together with one "circular" line.

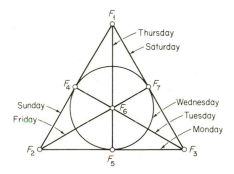

Figure 11.1

Even though an ordinary line cannot be drawn through the points F_4, F_5 and F_7, suppose we interpret this triple of points as a seventh line. Then we have seven points and seven lines such that every pair of points lies on a unique line and every pair of lines has a unique point of intersection. Viewed as a collection of points and lines, this system is called the **Fano plane**. Clearly the Fano plane incorporates all the properties required for constructing the firemen's timetable in Example 1. It is the mathematical abstraction which has this timetable as a possible *interpretation* or *application* or *realization*.

A projective plane consists of a *set of points* and a *set of lines* together with an *incidence relation* which allows us to state for every point and for every line either that the point is on the line or that the point is not on the line. In order to have a mathematical system called a *projective plane*, it is necessary to have certain rules or axioms satisfied. We may state all this more formally as follows.

A **projective plane** is a triple $(\mathscr{P}, \mathscr{L}, I)$ where $\mathscr{P}$ and $\mathscr{L}$ are sets (whose elements are called points and lines respectively) and I is a relation (called **incidence**) between points and lines which satisfies the following axioms.

PG1   Every line is incident with at least three points.
PG2   There exists at least one point and one line which are not incident.

PG3   Every pair of distinct lines is incident with a unique point.
PG4   Every pair of distinct points is incident with a unique line.

The usual terminology may be used. For example, PG3 could be stated as:

Every pair of distinct points lies on a unique line.

The first two axioms are *existence* axioms which are included to ensure that the systems defined are nontrivial. For example, PG2 ensures that all the points do not lie on a single line.

To illustrate the implications of these axioms, we prove the following theorem.

THEOREM 1   In a projective plane $(\mathscr{P}, \mathscr{L}, I)$ the set $\mathscr{P}$ has cardinality ≥ 7.

Proof   By PG2 there is a point P_1 not on some line L_1 (Figure 11.2). By PG1 there are at least three points P_2, P_3, and P_4 on L_1 (hence different from P_1). The pair of points P_1, P_2 must lie on a line L_2 which, by PG1, must contain a point P_5 different from P_1 and P_2. It follows that P_5 is different from P_3, for otherwise P_5 would lie on L_1. In the same way, it can be shown that $P_5 \neq P_4$. Similarly, there are two further points P_6 and P_7 on lines determined by P_1 and P_3 and by P_1 and P_4. Thus there are at least seven distinct points in the plane and the theorem is proved.

The Fano plane shown in Figure 11.1 is a projective plane having exactly seven points. This proves the following Corollary.

COROLLARY   There exists a projective plane with seven points.

The results stated in the following theorem can be proved from the axioms. The proof is left as an exercise.

THEOREM 2   Assume that there are $m + 1$ points on a *given* line L of a projective plane $(\mathscr{P}, \mathscr{L}, I)$. Then

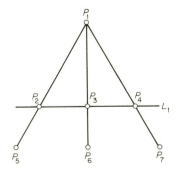

Figure 11.2

   (i)   *every* line of the projective plane contains $m + 1$ points;
   (ii)   every point of the projective plane lies on $m + 1$ lines;
   (iii) the cardinality of $\mathscr{P}$ is $m^2 + m + 1$;
   (iv)   the cardinality of $\mathscr{L}$ is $m^2 + m + 1$.

   In the next section it will be shown that projective geometries exist for every integer m of the form $m = p^n$ where p is a prime. It is known that there is no plane if $m = 6$, but it is not known whether a plane exists for $m = 10$. Many planes have been found for $m = 9$, but no plane has yet been found for which m is not a power of a prime. (For further information, see M. Hall, Jr. "Combinatorial Theory," Blaisdell, Waltham, Massachusetts, 1967.)

   The connection between projective geometry and ordinary geometry will be explained in the next section.

EXERCISE 11.2

1. Verify that the schedule shown in Table 11.2 has the property that every pair of firemen works together exactly one night each week.

2. Prove Theorem 2.

3.* Try to construct a projective plane with four points on each line.

4.* An **affine plane** is a triple ($\mathscr{P}$, $\mathscr{L}$, I) which satisfies the following axioms.

   AG1   Every pair of points is incident with a unique line.
   AG2   Every point P not on a line L lies on a unique line M which does not intersect L.
   AG3   There are at least four points, no three of which are collinear.

   (a) Show that an affine plane can be obtained from any projective plane by deleting one line and all the points on it.
   (b) State and prove a theorem analogous to Theorem 2 for an affine plane.

5.* Prove that axioms PG1, PG2 can be replaced by the single axiom AG3.

11.3 Coordinate Geometry

   In ordinary analytic geometry we represent points of the plane as ordered pairs (x, y) of real numbers and then lines are sets of points that satisfy real equations of the form $ax + by + c = 0$ with a and b not both zero. Since the algebra involved depends only on addition, subtraction, multiplication, and division, the field of real numbers can be replaced by any other field, in particular, a finite field. In this section, we discuss this type of geometry which is known as **affine geometry**, and we show how such a geometry can be extended to yield an analytic projective geometry.

We repeat here the definition of an affine plane given in Problem 4 of Exercise 11.2.

An **affine plane** is a triple $(\mathscr{P}, \mathscr{L}, I)$ which satisfies the following three axioms.

AG1 Every pair of points is incident with a unique line.

AG2 Every point P not on a line L lies on unique line M which does not intersect L.

AG3 There are at least four points in the plane, no three of which are collinear.

We now prove the following theorem.

THEOREM 1   Let $\mathscr{F} = (F, +, \cdot)$ be any field (in particular a finite field). Let $\mathscr{P}$ denote the set of ordered pairs (x, y) with x and y in F, and let $\mathscr{L}$ consist of those subsets L of $\mathscr{P}$ which satisfy linear equations; that is, $L \in \mathscr{L}$ if, for fixed $a, b, c, L = \{(x, y) \mid ax + by + c = 0, a, b, c \in F, a, b$ both not zero$\}$. A point $P \in \mathscr{P}$ is incident with (bears relation I to) a line $L \in \mathscr{L}$ if $P \in L$. Then $(\mathscr{P}, \mathscr{L}, I)$ is an affine plane which we denote $\mathscr{A}(\mathscr{F})$.

Proof The proof is straightforward. The points $P_1 = (x_1, y_1)$ and $P_2 = (x_2, y_2)$ determine the unique line

$$(y_2 - y_1)x - (x_2 - x_1)y = x_1 y_2 - x_2 y_1,$$

thus satisfying axiom AG1. If $P_1 = (x_1, y_1)$ is not on the line L determined by the equation $ax + by + c = 0$, then $ax_1 + by_1 + c \neq 0$ and

$$ax + by - (ax_1 + by_1) = 0$$

is a line L^* which contains P_1. It is easily verified that L and L^* have no point in common and that L^* is unique. Thus AG2 is satisfied. Finally, AG3 may be verified by noting that $(0, 0)$, $(0, 1)$, $(1, 0)$, and $(1, 1)$ are four points with no three collinear. By the definition, $(\mathscr{P}, \mathscr{L}, I)$ is an affine plane, as required.

One result that can readily be shown is that if $\mathscr{F} = (F, +, \cdot)$ and F contains m elements, then each line of $\mathscr{A}(\mathscr{F})$ contains exactly m points. The reader is asked to prove this in Problem 2 of Exercise 11.3.

In Problem 4 of Exercise 11.2, it was asserted that an affine plane can be constructed from a projective plane by deleting a line (and all the points on it).

We can reverse the process and construct a projective plane from $\mathscr{A}(\mathscr{F})$ by *adding* a line to it.

Let us start with $\mathscr{A}(\mathscr{F})$ and rename all the points as $(x, y, 1)$, that is, (x, y, z) with $z = 1$, and use the equation $ax + by + cz = 0$ (a, b, c not all zero) as the equation of a line, rather than $ax + by + c = 0$. It is clear that we have changed nothing in $\mathscr{A}(\mathscr{F})$ except the notation. Now add the additional set of points (the line L_∞),

$$L_\infty = \{(1, 0, 0), (x, 1, 0), x \in \mathscr{F}\}$$

to $\mathscr{P}$ to form a new set $\mathscr{P}' = \mathscr{P} \cup L_\infty$ (if $\mathscr{F}$ is a finite field containing m elements, then we have added $m + 1$ points to $\mathscr{P}$). Notice that the points of L_∞ can be represented by the equation $z = 0$ and so can be interpreted as a line.

Notice, also, that $(0, 0, 0)$ is not a point of the enlarged set. Let this new line L_∞ be added to $\mathscr{L}$ to form the set $\mathscr{L}' = \mathscr{L} \cup \{L_\infty\}$. With the natural extended notion of incidence, it can be verified that $(\mathscr{P}', \mathscr{L}', I')$ satisfies all the axioms for a projective plane. We denote this plane by $PG(\mathscr{F})$.

THEOREM 2   Let $\mathscr{A}(\mathscr{F}) = (\mathscr{P}, \mathscr{L}, I)$ and let

$$\mathscr{P}' = \mathscr{P} \cup \{(1, 0, 0), (x, 1, 0), x \in F\} = \mathscr{P} \cup L_\infty,$$
$$\mathscr{L}' = \mathscr{L} \cup \{L_\infty\},$$

and let the extended incidence relation be denoted by I'. Then $(\mathscr{P}', \mathscr{L}', I')$ is a projective plane $PG(\mathscr{F})$.

EXAMPLE 1   The plane $PG(GF(2))$, that is, the projective plane over the field $\mathscr{F}_2 = GF(2)$ containing two elements 0 and 1, has seven points: the four points $(0, 0, 1), (1, 0, 1), (0, 1, 1)$, and $(1, 1, 1)$ with $z \neq 0$, and the three distinct points on the line $z = 0$, namely $(0, 1, 0), (1, 0, 0)$, and $(1, 1, 0)$.

It can be verified that $PG(GF(2))$ also contains seven lines:

$$x = 0, \qquad y = 0, \qquad z = 0, \qquad x + y = 0,$$

$$x + z = 0, \qquad y + z = 0, \qquad \text{and} \qquad x + y + z = 0.$$

The seven points and lines are represented in Figure 11.3.

Comparing Figure 11.3 with Figure 11.1 of Section 11.2, it is clear that $PG(GF(2)) = PG(2)$ is the Fano plane. Note the use of the symbol $PG(m)$ to stand for the more awkward $PG(GF(m))$.

Now in constructing $PG(\mathscr{F})$, every line of $\mathscr{A}(\mathscr{F})$ must meet the new line L_∞ so there will be an additional point on each line; also the new line L_∞ contains

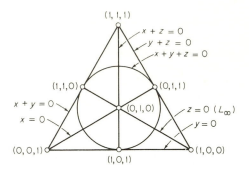

Figure 11.3

$m + 1$ points. Thus if $\mathscr{F}$ contains m elements, every line of $PG(\mathscr{F})$ contains $m + 1$ points.

Since there are finite fields for every $m = p^n$, p prime, we have the following theorem.

THEOREM 3   There exists a finite projective plane $PG(p^n)$ corresponding to every prime p and every positive integer n.

The additional line L_∞ added to an affine plane to obtain a projective plane is sometimes called the **line at infinity**. If two lines intersect on L_∞, they are called parallel. Thus theorems in projective geometry will hold in affine geometry if we interpret lines which intersect on L_∞ as parallel lines.

EXAMPLE 2   In Example 1, the line at infinity, L_∞, is the line $z = 0$ containing the points $(1, 1, 0)$, $(0, 1, 0)$, and $(1, 0, 0)$. Then the lines $x + z = 0$ and $x = 0$ are parallel because they intersect in the point $(0, 1, 0)$ on L_∞. Note also that $GF(2)$ contains two elements and every line of $PG(GF(2))$ contains $2 + 1$ points.

The above interpretation is particularly useful in case $\mathscr{F}$ is the field $\mathscr{R}$ of real numbers. Then the above remarks indicate that all the theorems which hold in the real projective plane can be reinterpreted for ordinary Euclidean plane geometry. Since affine and projective geometries are based on incidence, such theorems do not involve measurements of distance or angle. Examples of projective theorems are presented in the next section.

It is interesting to note that there exist projective planes other than the ones discussed in this section. These projective planes *cannot* be represented analytically in terms of fields. They are called **non-desarguesian planes**, since Desargues's theorem (to be discussed in Section 11.4) does not hold.

EXERCISE 11.3

1.   Enumerate the points and lines of $\mathscr{A}(GF(3))$.

2.   Show that if $\mathscr{F} = (F, +, \cdot)$ and F contains m elements, then each line of $\mathscr{A}(\mathscr{F})$ contains exactly m points.

3.   Complete the proof of Theorem 2.

4.   (a)   In Example 2, enumerate *all* sets (pencils) of pairs of parallel lines.
    (b)   Point $(0, 1, 1)$ is not on the line $y = 0$. Is there a line through $(0, 1, 1)$ which is parallel to the line $y = 0$?
    (c)   Repeat (b) for the point $(0, 0, 1)$ and the line $y + z = 0$.

5.* (a)   Construct the points and lines of $PG(3)$. Draw a diagram showing all the intersections.
    (b)   Enumerate the points on L_∞ and the families of parallel lines.

6.* Let $S = \{(x, y, z); x, y, z \in F\}$ and let (x, y, z) and (x', y', z') represent the same point if the coordinates are proportional. Let $\mathscr{P}^*$ denote the resulting set of points. Similarly let lines in $\mathscr{L}^*$ be sets of points which satisfy the same linear equations $ax + by + cz = 0$ (not all $a, b, c,$ equal to zero). The equation $kax + kby + kcz = 0$ represents the same line for all $k \neq 0$. Incidence I^* is defined in the natural way. Prove that $(\mathscr{P}^*, \mathscr{L}^*, I^*)$ is the projective plane $PG(\mathscr{F})$. (Coordinates defined in this way are called **homogeneous coordinates**.)

11.4 Projective Configurations

In this section we present without proof two interesting theorems which hold in all projective planes that can be represented analytically in terms of fields, and a theorem which holds only in some of these projective planes.

Two triangles, $\triangle ABC$ and $\triangle A'B'C'$ are said to be **in perspective from a point** O if AA', BB', and CC' pass through O. In the following theorem we use the notation $BC \cdot B'C'$ to indicate the point of intersection of the lines BC and $B'C'$.

THEOREM 1 (Desargues's Theorem)   If $\triangle ABC$ and $\triangle A'B'C'$ are in perspective, then the points $P = BC \cdot B'C'$, $Q = CA \cdot C'A'$, and $R = AB \cdot A'B'$ are collinear.

The theorem is illustrated in Figure 11.4 and an analytic proof is suggested as an exercise (Problem 2 of Exercise 11.4). The reader can consult any book on projective geometry for further information.

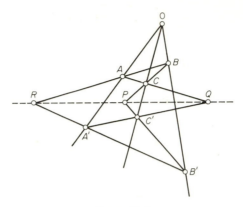

Figure 11.4

THEOREM 2   (The Theorem of Pappus)   Let A, B, and C denote points on a line L and let A', B', and C' denote points on a second line L'. Let $P = BC' \cdot B'C$, $Q = CA' \cdot C'A$, and $R = AB' \cdot A'B$. Then P, Q, and R are collinear.

The diagram in Figure 11.5 illustrates this theorem.

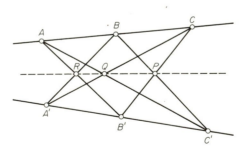

Figure 11.5

Theorems 1 and 2 play an important role in projective geometry. If Desargues's theorem holds, then coordinates can be defined (as in Section 11.3) in terms of elements from a division ring. A **division ring** is like a field except that the commutative law need not hold. If the theorem of Pappus holds, then the multiplication in the division ring is necessarily commutative; that is, $ab = ba$ for all a, b and so the division ring is actually a field. Since it can be shown that $ab = ba$ in every *finite* division ring, that is, the only finite division rings are fields, then the theorem of **Pappus** holds in $PG(\mathscr{F})$ where $\mathscr{F}$ is any finite field.

As a final example, we present a theorem which holds only in certain projective planes, and does not hold, for example, in the real projective plane.

THEOREM 3   Let A, B, C, and D denote four points, no three collinear, of the projective plane $PG(\mathscr{F})$. Let $P = AB \cdot CD$, $Q = AC \cdot BD$, and $R = AD \cdot BC$. Then P, Q, and R are collinear in $PG(m)$, if and only if $m = 2^n$ for some n, that is, if and only if the characteristic of the field $\mathscr{F}$ is 2.

This situation is illustrated in Figure 11.6 where the dotted curve is to be interpreted as a line. It is impossible to choose the points A, B, C, and D in the real plane so that P, Q, and R are collinear since P, Q, and R are the three diagonal points of the quadrangle $ABCD$. The proof of the theorem is requested in Problem 4 of Exercise 11.4.

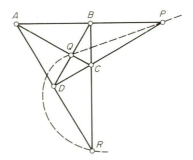

Figure 11.6

EXERCISE 11.4

1. Convince yourself that Desargues's theorem holds in the real plane by drawing a number of diagrams with the aid of a ruler.

2.* Let the coordinates of A, B, C, D be taken as $(1, 0, 0)$, $(0, 1, 0)$, $(0, 0, 1)$, $(1, 1, 1)$ respectively. Use homogeneous coordinates (Problem 6 of Exercise 11.3) to obtain the coordinates of P, Q, R of Figure 11.4 in terms of the coordinates of A', B', C'. Hence prove Desargues's theorem.

3.* Prove the theorem of Pappus by judiciously choosing the coordinate axes $x = 0$, $y = 0$, and $z = 0$. The point $(1, 1, 1)$, the unit point, can be taken as any point not on the coordinate triangle.

4.* Prove Theorem 3.

III

APPLICATIONS

The enumeration methods of Part I and the methods of proof of Part II have applications to combinatorial problems as well as to areas of applied mathematics dealing with finite sets of objects. In Part III we consider some of these applications. Chapter 12 considers problems involving probability and Chapter 13 considers generalizations of the binomial coefficients, including an original section on arithmetic power series. In Chapter 14 we discuss further applications of generating functions and difference equations, applying these techniques to the triangulation of polygons, further work on random walks, and ideas associated with mathematical operations research. Chapter 15 presents some work on the famous Fibonacci sequence, and in Chapter 16 we include some interesting examples of combinatorial structures.

Probability

Suppose that in a given situation, such as the performance of an experiment, the play of a game or contest, or a real life situation, etc., there are n possible outcomes or results $\theta_1, \theta_2, \ldots, \theta_n$, which are mutually exclusive. We may assign a nonnegative real number $p_i = p(\theta_i)$, $i = 1, 2, \ldots, n$, with

$$p_1 + p_2 + \cdots + p_n = 1,$$

to each of these possible outcomes. These real numbers represent an estimate of the relative likelihood of the different outcomes. This assignment of the nonnegative real numbers p_i to the outcomes θ_i is not a mathematical problem, but once the assignment has been made, the resulting problems *are* mathematical. If an event E occurs which has any one of the k possible outcomes $\theta_{i_1}, \theta_{i_2}, \ldots, \theta_{i_k}$, then to this event E we assign the **probability** $\Pr(E) = p_{i_1} + p_{i_2} + \cdots + p_{i_k}$. In many practical situations it is natural to consider a number of outcomes as equally likely. If the n possible outcomes $\theta_1, \theta_2, \ldots, \theta_n$ are equally likely, then $p_1 = p_2 = \cdots = p_n = 1/n$ and in this case, if an event E occurs with any one of k possible outcomes, then $\Pr(E) = k/n$. In such cases, the calculation of the probability $\Pr(E)$ of an event E is equivalent to the combinatorial problem of calculating k, the number of possible outcomes yielding the event E.

12.1 Combinatorial Probability

In probability, the elements or points of the *sample space* of an experiment are the various possible results or outcomes of the experiment.

EXAMPLE 1   If an experiment consists of tossing three different coins, say a nickel, dime, and quarter, the sample space is the set of ordered triples.

{HHH, HHT, HTH, HTT, THH, THT, TTH, TTT}

where the entries H and T designate heads and tails respectively for the three coins. If the three coins tossed were not distinguishable, the sample space would be the set of unordered triples

{HHH, HHT, HTT, TTT}.

As described in the introduction, a certain numerical weight, called a probability, is associated with each element in the sample space. In the experiment where three distinguishable coins are tossed, all events are usually assigned equal probabilities of $\frac{1}{8}$. In general, these weights or probabilities are attached to the elements of the sample space in such a way that the sum of the probabilities for all elements of the sample space is 1. With this convention, and the additional convention that the probability associated with an element cannot be negative, we see that each element of the sample space has attached to it a number p with $0 \le p \le 1$.

An **event** A is defined to be a subset of the sample space; in the case of tossing three distinguishable coins, various possible events are:

(a)   three heads (only one element of the sample space is in this event);
(b)   any result (the entire sample space);
(c)   one head (this event contains the three elements HTT, THT, TTH);
(d)   at least two heads (this event contains the four elements HHH, HHT, HTH, THH).

From the definition of probability, the probability of an event A is defined to be the sum of the probabilities of the elements in A. Thus the probability of an event A (or the " weight " of the set A) is defined as some number between 0 and 1. In the example above, the probabilities in (a), (b), (c), and (d) are $\frac{1}{8}$, 1, $\frac{3}{8}$, and $\frac{1}{2}$ respectively.

Similarly, when the sample space consists of three indistinguishable (unordered) coins as in Example 1, it is natural to associate the probabilities $\frac{1}{8}$, $\frac{3}{8}$, $\frac{3}{8}$, $\frac{1}{8}$ with the elements, consistent with the various orderings of the coins in the sample space of distinguishable (ordered) coins.

Suppose an event consists of selecting one of a set of objects, each of which is equally likely to be chosen. Then if there are several types of objects, the probability of selecting an object of a given type is proportional to the number of that type of object in the whole collection.

THEOREM   Suppose N objects are of k types, $t_1, t_2, \ldots, t_k$, and the number of objects of type t_i is N_i, $i = 1, 2, \ldots, k$. If it is equally likely that a given object is any one of the N objects, the probability that a given object is of type t_i is $p(t_i) = N_i/N$, $i = 1, 2, \ldots, k$.

The above statement could be taken as an axiom rather than a theorem, depending on the approach we may wish to take.

EXAMPLE 2   The cards of an ordinary 52-card deck are shuffled; what is the probability that in the resulting arrangement the ace and king of diamonds are together?

Solution   Our sample space consists of all 52! possible arrangements of the 52 cards. To each of these elements of the sample space, we assign equal weights or probabilities; namely, the positive real number 1/52!

If we consider the ace and king of diamonds as glued together, there will then be 51 "cards" to arrange, and so there will be 51! arrangements of the "cards" in which the ace of diamonds is on top of the king, and another 51! arrangements in which the king of diamonds is on top of the ace. Hence the required probability is

$$\frac{2(51!)}{52!} = \frac{1}{26}.$$

The probability of A is often denoted by $\Pr(A)$; then $\Pr(A \cap B)$ or $\Pr(AB)$ denotes the probability of both events A and B occurring, and $\Pr(A \cup B)$ denotes the probability of at least one of the events A and B occurring. From our work on sets (Section 3.2), we know that

$$n(A \cup B) = n(A) + n(B) - n(AB)$$

where $n(A)$ denotes the number of elements in the set A. Now each element of the sample space is assigned equal probability, that is, the probability of any event, such as A, is proportional to the number $n(A)$ of elements in A. Hence

$$\Pr(A \cup B) = \Pr(A) + \Pr(B) - \Pr(AB).$$

Another way of looking at this is as follows. If $\mathscr{U}$ denotes the sample space of all possible events, then $\Pr(A) = n(A)/n(\mathscr{U})$, with similar expressions for

Pr(B), Pr(AB), and Pr($A \cup B$).) Thus dividing the numbers of the above expression for $n(A \cup B)$ by $n(\mathcal{U})$ yields the above equation relating the probabilities.

EXAMPLE 3   Let a sample space be made up of all ($^{52}_{5}$) possible five-card hands drawn from an ordinary deck of cards. Find the probability that a five-card hand contains either the ace or king of hearts.

Solution   Let A be the event that the hand contains the ace of hearts and B the event that the hand contains the king of hearts. Then

$$n(A) = \text{number of elements in set } A = \binom{51}{4}$$

$$= n(B)$$

$$n(AB) = \binom{50}{3}.$$

If p is the probability attached to any element in the sample space, then $p = 1/\binom{52}{5}$ and

$$\mathrm{Pr}(A \cup B) = \mathrm{Pr}(A) + \mathrm{Pr}(B) - \mathrm{Pr}(AB)$$

$$= \binom{51}{4}p + \binom{51}{4}p - \binom{50}{3}p$$

$$= \frac{5}{52} + \frac{5}{52} - \frac{5 \cdot 4}{52 \cdot 51} = \frac{5 \cdot 49}{51 \cdot 26}.$$

We have analogous results for the case of more than two events. In particular, a most useful expansion is

$$\mathrm{Pr}(A \cup B \cup C) = \mathrm{Pr}(A) + \mathrm{Pr}(B) + \mathrm{Pr}(C)$$
$$- \mathrm{Pr}(AB) - \mathrm{Pr}(BC) - \mathrm{Pr}(CA) + \mathrm{Pr}(ABC).$$

This follows from the expression for $n(A \cup B \cup C)$ developed in Section 3.2.

As in the case of the inclusion–exclusion theorem, this generalizes to the following theorem which is stated without proof.

THEOREM   Let $A, B, \ldots, N$ denote n events (not necessarily independent). Then

$$\mathrm{Pr}(A \cup B \cup \cdots \cup N) = \mathrm{Pr}(A) + \mathrm{Pr}(B) + \cdots + \mathrm{Pr}(N)$$
$$- [\mathrm{Pr}(AB) + \mathrm{Pr}(AC) + \cdots]$$
$$+ [\mathrm{Pr}(ABC) + \cdots]$$
$$- \cdots + (-1)^{n-1}\mathrm{Pr}(ABC \cdots N).$$

EXAMPLE 4   What is the probability that a hand of thirteen cards dealt at random will contain (a) exactly one ace, (b) at least one ace?

Solution   Let A be the event that a hand of thirteen cards contains exactly one ace and B the event that such a hand contains at least one ace. Then

$$\Pr(A) = \frac{\binom{4}{1}\binom{48}{12}}{\binom{52}{13}}$$

and

$$\Pr(B) = 1 - (\text{probability that the hand contains no aces})$$

$$= 1 - \frac{\binom{48}{13}}{\binom{52}{13}}.$$

The reader should translate each of the above expressions into words. For example, $\binom{4}{1}\binom{48}{12}$ is the number of hands containing one ace, since an ace may be chosen from four aces in $\binom{4}{1}$ ways and the remaining 12 cards of the hand may be chosen in $\binom{48}{12}$ ways from the 48 cards which are not aces. The reader should similarly interpret the expressions in Example 5.

EXAMPLE 5   What is the probability that a hand of thirteen cards dealt at random contains (a) exactly two diamonds, (b) the ace of hearts and a void in clubs, (c) the ace of hearts and twelve black cards?

Solution   Let A, B, and C represent the events described in (a), (b), and (c) respectively. Then

$$\Pr(A) = \frac{\binom{13}{2}\binom{39}{11}}{\binom{52}{13}}, \qquad \Pr(B) = \frac{\binom{38}{12}}{\binom{52}{13}}, \qquad \Pr(C) = \frac{\binom{26}{12}}{\binom{52}{13}}.$$

The following exercises are of an elementary nature in which we assume that all points of the sample space have equal probabilities.

EXERCISE 12.1

1.   A man has eight pairs of socks in a drawer in a dark room; he takes four socks at random into another room which is light. Find the probability that (a) he will not have a matching pair, (b) he will have at least one matching pair. (Assume all eight pairs are different in color.)

2.   Consider a bridge deck and let the sample space consist of all possible drawings of a card from the deck. Then the sample space consists of 52 elements. Let events A, B, and C be defined respectively as drawing

a spade, drawing a six, and drawing an honor card (A, K, Q, J, 10). Determine the probabilities for the events A, B, C, AB, BC, CA, and ABC, and find $\Pr(A \cup B \cup C)$, $\Pr(A\beta C)$, $\Pr(A\beta\gamma)$, and $\Pr(\alpha\beta)$. Note that $\alpha = $ "not A" $= A'$, $\beta = $ "not B" $= B'$, and $\gamma = $ "not C" $= C'$.

3. Take the digits from zero to nine and let the sample space consist of all possible ordered choices of five distinct digits. Then the sample space contains $10 \cdot 9 \cdot 8 \cdot 7 \cdot 6$ elements. Let A be the event that the selection ends in an even digit, B the event that no one of the five digits exceeds seven. Calculate the probabilities $\Pr(A)$, $\Pr(B)$, $\Pr(AB)$, and $\Pr(\alpha\beta)$.

4. Consider a bridge deck and let the sample space be the drawing of a hand of thirteen cards. Find the probability that no void occurs in the hand. (Problems of this kind are very simple if we clearly state what the sample space and the A_i are; here we let A_i be the event the hand contains no cards of suit i.)

5. A coin is tossed seven times; find the probability of a run of at least four successive tails.

6.* Generalize Problem 5 to consider the case of a run of length at least r successive tails in n tossings, given that $2r \geq n$.

7. Find the probability that a random bridge hand contains (a) exactly three hearts, (b) the ace and king of clubs, (c) the ace and king of clubs and eleven red cards.

8. Find the probability that a hand of thirteen cards, dealt at random, will contain (a) exactly four black cards, (b) no black cards.

9. Find the probability that a random bridge hand consists of four hearts, four spades, three diamonds, and two clubs.

10.* (a) Twenty cards are drawn from a deck; find the probability that the entire spade suit is included.

   (b) Generalize (a) to the case of a sample of r cards drawn from a deck of n cards with the condition that the sample is to include a suit of b cards.

12.2 Ultimate Sets

In many applications of probability there are events which are not independent, for example, the events A and B of Example 4 of Section 12.1; the hands containing exactly one ace also classify as hands containing at least one ace so that event B contains event A. This happens when we are dealing with objects that are classified under several different attributes or properties simultaneously. The sets of objects that correspond to independent events are ultimate sets. We proceed to define what we mean by ultimate sets.

Let $A, B, C, \ldots, N$ denote n different subsets of a universal set U and let $\alpha, \beta, \gamma, \ldots, \nu$ denote their complements $A', B', C', \ldots, N'$ in U. There are 2^n sets that can be formed by taking intersections of n of the $2n$ sets $A, B, \ldots, N, \alpha, \beta, \ldots, \nu$ under the proviso that no set occur with its complement in a product. Some of the sets formed in this way may be the same in special cases but in general we obtain 2^n formal sets with the given property. We call these 2^n formal sets the **ultimate sets** associated with the n given subsets of U and their complements $\alpha, \beta, \ldots, \nu$ in U.

EXAMPLE 1   If $n = 2$ we have two different subsets A and B of U and two complements α and β. The ultimate sets are

$$AB, \quad A\beta, \quad \alpha B, \quad \text{and} \quad \alpha\beta.$$

EXAMPLE 2   If $n = 3$, we have three different subsets A, B, and C of U and three complements α, β, γ. The eight ultimate sets are

$$ABC, \quad AB\gamma, \quad A\beta C, \quad A\beta\gamma, \quad \alpha BC, \quad \alpha B\gamma, \quad \alpha\beta C, \quad \alpha\beta\gamma.$$

It should be clear to the reader that ultimate sets are mutually disjoint. Any of the $2n$ given sets can be expressed as a sum of ultimate sets.

EXAMPLE 3   If $n = 2$, $A = AU = A(B + \beta) = AB + A\beta$.
If $n = 3$, $A = AUU = A(B + \beta)(C + \gamma)$
$$= ABC + AB\gamma + A\beta C + A\beta\gamma$$
$$\beta = U\beta U = (A + \alpha)\beta(C + \gamma)$$
$$= A\beta C + A\beta\gamma + \alpha\beta C + \alpha\beta\gamma.$$

EXAMPLE 4   Express $\beta\gamma$ in terms of ultimate sets related to subsets $A, B, C,$ and D of U.

Solution

$$\beta\gamma = U\beta\gamma U$$
$$= (A + \alpha)\beta\gamma(D + \delta)$$
$$= A\beta\gamma D + A\beta\gamma\delta + \alpha\beta\gamma D + \alpha\beta\gamma\delta.$$

These examples suggest that every set that can be formed from subsets $A, B, C, \ldots$ of U by the usual set operations of union, intersection, and complementation, can be expressed as a sum of ultimate sets. For every such set is a sum of products involving the sets $A, B, \ldots,$ and their complements, and any product can be expressed in terms of ultimate sets by introducing factors $X + X' (=U)$ corresponding to any of the sets $A, B, \ldots$ which, along with their complements, are not contained in the product, and then "multiplying out" the result. This was illustrated in Examples 3 and 4. Thus we have the following theorem.

THEOREM 1   Let S denote any set which can be formed from the different subsets $A, B, C, \ldots$ of a universal set U by set operations. Then S can be expressed as a sum of the ultimate sets determined by $A, B, C, \ldots$.

Suppose probabilities have been associated with the subsets of U. In cases where U has been given some interpretation, this is easy to do. For example, suppose U is the set of all 1963 cars still on the road. Then A could be the subset of blue cars, B the subset of Chevrolets, etc. The probability that a car is in a set X is proportional to the number of elements in it. Thus if probabilities have been associated with the ultimate sets, the probability associated with a given set can be deduced by invoking the following Theorem 2.

THEOREM 2   Let the sets $A, B, C, \ldots$ determine ultimate sets $U_1, U_2, \ldots, U_k$ with associated probabilities $p_1, p_2, \ldots, p_k$. Suppose an event E determines a subset S which is expressible in terms of $A, B, C, \ldots$. Then

$$S = U_{i_1} + U_{i_2} + \cdots + U_{i_k}$$

and

$$\Pr(E) = p_{i_1} + p_{i_2} + \cdots + p_{i_k}.$$

Since the ultimate sets are, in a sense, the building blocks of the subsets of a set U, then the ultimate sets cannot have negative cardinality otherwise we may conclude that there is something wrong with our work. This idea may be applied to test for consistency given data involving various characteristics. This is illustrated in Example 5.

EXAMPLE 5   Check the following data for consistency. From a class of 23, 10 passed math, 10 passed French, 14 passed English, 8 passed math and French, 7 passed math and English, 4 passed English and French, and 3 passed all three.

Solution   Let $A =$ set of math passers
$B =$ set of French passers
$C =$ set of English passers.

Then $n(U) = 23$, $n(A) = n(B) = 10$, $n(C) = 14$, $n(AB) = 8$, $n(AC) = 7$,
$n(BC) = 4$, $n(ABC) = 3$. Hence,

$n(ABC) = 3$

$AB\gamma = AB(U - C) = AB - ABC,$	$n(AB\gamma) = 5$
$\alpha BC = BC - ABC,$	$n(\alpha BC) = 1$
$A\beta C = AC - ABC,$	$n(A\beta C) = 4$
$\alpha\beta C = (U - A)(U - B)C = C - BC - AC + ABC,$	$n(\alpha\beta C) = 6$
$\alpha B\gamma = (U - A)B(U - C) = B - AB - BC + ABC,$	$n(\alpha B\gamma) = 1$
$\alpha\beta\gamma = U - A - B - C + AB + AC + BC - ABC,$	$n(\alpha\beta\gamma) = 5$
$A\beta\gamma = A(U - B)(U - C) = A - AB - AC + ABC,$	$n(A\beta\gamma) = -2.$

$n(A\beta\gamma)$ is negative so that the data are inconsistent.

E X E R C I S E 12.2

1. Express (a) $\alpha\beta\gamma$, (b) αCD, (c) βD, (d) α, (e) γ in terms of ultimate sets related to A, B, C, D.

2. Check the following data for consistency.

From a group of 100 students surveyed, 62 were undergraduates, 38 were married, 69 had cars, 28 were married undergraduates, 30 undergraduates had cars, 29 married students had cars, and 18 married undergraduates had cars.

3. Check the following data for consistency.

A recent survey of 1000 children reports that 730 were boys, 600 used Summit toothpaste, and 650 had no cavities. In addition, 450 boys used Summit, 475 boys had no cavities, 375 children who used Summit had no cavities, and 325 boys who used Summit had no cavities.

4.* If, given n sets, A, B, C, ..., N, no set is contained in the union of all the others, the n sets are said to be *independent*. Prove that if no ultimate set related to these n sets is empty, the given sets are independent.

Ramifications of the Binomial Theorem

There are many applications and extensions of the binomial theorem. A few of these have already been mentioned. For example, the multinomial theorem of Section 2.6 is an extension of the binomial theorem and the formula for the number of paths from the origin to the point $(n - r, r)$ is an application of the binomial coefficients (Section 2.5). In this chapter we discuss several other ways in which binomial-type expressions occur. In Section 13.1 we present some work by Kenneth Berman in which he applies the binomial theorem to arithmetic power series. Section 13.2 involves a connection between the binomial theorem and probability, and Section 13.3 involves the binomial theorem in a discussion of the distribution of objects into boxes. In Section 13.4 we introduce Stirling numbers which are useful whenever products of the form $n^{(r)}$ occur, for example in determining chromatic polynomials. Finally, in Section 13.5, we present a generalization of binomial coefficients first studied by Gauss.

13.1 Arithmetic Power Series

In this section we will develop a recursion formula for summing an arithmetic power series, a series of the form

$$S_n^{\ k} = a^k + (a + d)^k + (a + 2d)^k + \cdots + (a + nd)^k \tag{1}$$

in which the terms on the right which are raised to the kth power are just the terms of the arithmetic series

$$S_n = a + (a + d) + (a + 2d) + \cdots + (a + nd).$$

This development illustrates at least four important techniques: (i) the symbolic approach, (ii) the use of the binomial theorem, (iii) the use of recursion relations, and (iv) the summation method.

THEOREM   Let a, d, n, k, and x be positive integers and let

$$S_n^{\ k} = \sum_{i=0}^{n} (a + id)^k.$$

Then

$$(S_n + d)^x - S_n^{\ x} = [d(n + 1) + a]^x - d^x n - a^x$$

where $(S_n + d)^x$ is defined symbolically by

$$(S_n + d)^x = \binom{x}{0} S_n^{\ x} + \binom{x}{1} S_n^{\ x-1} d + \cdots + \binom{x}{x-1} S_n^{\ 1} d^{x-1} + \binom{x}{x} d^x,$$

that is, $(S_n + d)^x$ is expanded using the binomial theorem and the "powers" of S_n are regarded as being defined as above.

Proof   By the binomial theorem,

$$[(a + kd) + d]^x = \binom{x}{0}(a + kd)^x + \binom{x}{1}(a + kd)^{x-1} d + \cdots$$
$$+ \binom{x}{x-1}(a + kd)d^{x-1} + \binom{x}{x} d^x.$$

Now set $k = 0, 1, 2, \ldots, n$ in turn and add the corresponding members of the resulting equations. This yields

$$[a + d]^x + [(a + d) + d]^x + \cdots + [(a + nd) + d]^x$$
$$= \binom{x}{0} S_n^{\ x} + \binom{x}{1} S_n^{\ x-1} d + \cdots + \binom{x}{x-1} S_n^{\ 1} d^{x-1} + \binom{x}{x} d^x + nd^x$$
$$= (S_n + d)^x + nd^x,$$

by definition. The left side of this equation can be rewritten as

$$[a^x + (a + d)^x + (a + 2d)^x + \cdots + (a + nd)^x] + [(n + 1)d + a]^x - a^x$$
$$= S_n^{\ x} + [(n + 1)d + a]^x - a^x.$$

Therefore,

$$(S_n + d)^x + nd^x = S_n^{\ x} + [(n + 1)d + a]^x - a^x$$

and the theorem follows.

COROLLARY   If $a = 0$, $d = 1$, and $k = x$ in (1), then

$$S_n{}^x = 1^x + 2^x + 3^x + \cdots + n^x$$

and by the theorem,

$$(S_n + 1)^x - S_n{}^x = (n + 1)^x - n.$$

We illustrate the use of the theorem with a few examples.

EXAMPLE 1   Develop a formula for the sum of the series

$$\sum_{i=0}^{n} (1 + 2i).$$

Solution

$$\sum_{i=0}^{n} (1 + 2i) = 1 + (1 + 2) + (1 + 4) + \cdots + (1 + 2n)$$
$$= S_n{}^1 \quad \text{with } a = 1 \quad \text{and} \quad d = 2.$$

Set $a = 1$, $d = 2$, and $x = 2$ in the theorem. Then

$$(S_n + 2)^2 - S_n{}^2 = [2(n + 1) + 1]^2 - 2^2 n - 1$$
$$= 4S_n{}^1 + 4.$$

Note also that

$$(S_n + 2)^2 = \binom{2}{0} S_n{}^2 + \binom{2}{1} S_n{}^1 \cdot 2 + \binom{2}{2} \cdot 2^2$$
$$= S_n{}^2 + 4S_n{}^1 + 4.$$

Therefore

$$S_n{}^1 = \frac{(2n + 3)^2 - 4n - 5}{4} = (n + 1)^2.$$

EXAMPLE 2   Develop a formula for the sum of the general arithmetic series

$$\sum_{i=0}^{n} (a + id).$$

Solution

$$\sum_{i=0}^{n} (a + id) = S_n{}^1.$$

From the theorem with $x = 2$,

$$(S_n + d)^2 - S_n{}^2 = 2dS_n{}^1 + d^2$$
$$= [d(n + 1) + a]^2 - d^2 n - a^2.$$

Therefore,

$$S_n^1 = \frac{[d(n+1)+a]^2 - d^2n - a^2 - d^2}{2d}$$

$$= \frac{(n+1)}{2}(2a + nd).$$

EXAMPLE 3   Develop a formula for the sum

$$\sum_{i=0}^{n}(1 + 3i)^2.$$

Solution

$$\sum_{i=0}^{n}(1 + 3i)^2 = S_n^2 \quad \text{with} \quad a = 1, d = 3.$$

Set $a = 1$ and $d = 3$ in the theorem and then set $x = 3$ and $x = 2$ in turn in the theorem to obtain the equations

$$(S_n + 3)^3 - S_n^3 = 9S_n^2 + 27S_n^1 + 27 = [3(n+1) + 1]^3 - 3^3n - 1^3$$
$$(S_n + 3)^2 - S_n^2 = \qquad\qquad 6S_n^1 + 9 = [3(n+1) + 1]^2 - 3^2n - 1^2.$$

Subtract $\frac{9}{2}$ of the members of the second equation from the corresponding members of the first.

$$9S_n^2 - \tfrac{27}{2} = [(3n+4)^3 - 27n - 1] - \tfrac{9}{2}[(3n+4)^2 - 9n - 1].$$

Solving we find that

$$S_n^2 = \frac{6n^3 + 15n^2 + 11n + 2}{2}$$

$$= \frac{(n+1)(6n^2 + 9n + 2)}{2}.$$

EXERCISE 13.1

1. Develop a formula for the sums

    (a) $\displaystyle\sum_{i=0}^{n}(1 + 3i)$ (b) $\displaystyle\sum_{i=0}^{n}(1 + 5i).$

    and check your work with the formula in Example 2.

2. Develop a formula for the sum

    $$\sum_{i=0}^{n}(1 + 2i)^2.$$

3.   Develop a formula for the sum

$$\sum_{i=0}^{n} (a + id)^2.$$

and use this formula to check Problem 2 and Example 3.

4.* Use the corollary to the theorem of this section to find a formula for $\sum_{i=1}^{n} i^3$.

13.2 The Binomial Distribution

A set of probabilities $\{p_0, p_1, \ldots, p_n\}$ with $\sum_{i=0}^{n} p_i = 1$ is called a **probability distribution** or **frequency distribution**. Such a distribution may be represented by ultimate sets of a sample space. These sets are mutually disjoint and their sum is the universal set. If the probabilities are proportional to the number of elements, that is, each element is equally likely, then

$$p_i = \frac{\text{cardinality of } i\text{th ultimate set}}{\text{cardinality of universal set}}$$

and $\sum p_i = 1$.

An important class of probability distributions is related to the binomial theorem.

EXAMPLE 1   When a fair coin is tossed, the probability p_0 that if falls heads is $\frac{1}{2}$ and the probability p_1 that it falls tails is $\frac{1}{2}$ and $p_0 + p_1 = 1$.

If the coin is tossed twice, there are three possible events: two heads, a head and a tail, and two tails. These events have probabilities $p_0 = (\frac{1}{2})^2$, $p_1 = 2(\frac{1}{2})^2$, and $p_2 = (\frac{1}{2})^2$ respectively and $p_0 + p_1 + p_2 = 1$. If the coin is tossed three times, there are four events corresponding to no heads, one head, two heads, and three heads with probabilities $p_0 = (\frac{1}{2})^3$, $p_1 = 3(\frac{1}{2})^3$, $p_2 = 3(\frac{1}{2})^3$, and $p_3 = (\frac{1}{2})^3$, respectively, and again $p_0 + p_1 + p_2 + p_3 = 1$.

The pattern is becoming clear, and it can be shown by induction that if the coin is tossed n times, the probability that there will be exactly k heads is $p_k = \binom{n}{k}(\frac{1}{2})^n$. Note that

$$\sum_{k=0}^{n} p_k = \sum_{k=0}^{n} \binom{n}{k}\left(\frac{1}{2}\right)^n = \left(\frac{1}{2}\right)^n \sum_{k=0}^{n} \binom{n}{k}$$

$$= \left(\frac{1}{2}\right)^n \cdot 2^n = 1,$$

so that the probabilities p_i in this case form a binomial distribution.

If the probabilities of heads and tails in such a coin tossing are p and q, respectively, with $p + q = 1$, the probabilities p_k of exactly k heads in n tosses yield the general binomial distribution. More generally, consider any two mutually exclusive events E_1 and E_2 with probabilities p and q respectively, $p + q = 1$, and suppose we consider n trials (analogous to flipping a coin n times) in which either the event E_1 or the event E_2 can occur. Then the probability p_k that E_1 occurs in exactly k of the n trials is given by

$$p_k = \binom{n}{k} p^k q^{n-k}. \tag{1}$$

The set of probabilities $\{p_0, p_1, \ldots, p_n\}$ defined by (1) determines a probability distribution since

$$\sum_{k=0}^{n} p_k = \sum_{k=0}^{n} \binom{n}{k} p^k q^{n-k} = (q + p)^n = 1.$$

This probability distribution is called the **binomial distribution**.

The generating function f corresponding to the probability distribution $\{p_0, p_1, \ldots, p_n\}$ is given by $\sum_i p_i x^i$ (Chapter 6). Thus the generating function of the binomial distribution is

$$f(x) = (q + px)^n. \tag{2}$$

The above ideas may be generalized to r independent or mutually exclusive events $E_1, E_2, \ldots, E_r$ which occur with probabilities $p_1, p_2, \ldots, p_r$, respectively, $\sum_{i=1}^{r} p_i = 1$. Let $p_{j_1 j_2 \cdots j_r}$ denote the probability that in n trials the ith event occurs j_i times; then $\sum j_i = n$ and

$$p_{j_1 j_2 \cdots j_r} = \binom{n}{j_1, j_2, \ldots, j_r} p_1^{j_1} p_2^{j_2} \cdots p_r^{j_r}.$$

This yields the multinomial distribution. A generating function for the multinomial distribution is

$$f(x_1, x_2, \ldots, x_r) = (p_1 x_1 + p_2 x_2 + \cdots + p_r x_r)^n.$$

EXERCISE 13.2

1. Complete the inductive proof that the probability of exactly k heads in n tossings of a fair coin (Example 1) is

$$p_k = \binom{n}{k} \left(\frac{1}{2}\right)^n.$$

2.* The expected number of heads in Problem 1 is defined by

$$\mu = p_1 + 2p_2 + \cdots + np_n.$$

Find μ (a) by summing the series defining μ, (b) by using the observation of Section 6.5 that $\mu = f'(1)$ where f is the generating function of the distribution $\{p_1, p_2, \ldots, p_n\}$.

3.* Evaluate μ, defined in Problem 2, if the distribution is the binomial distribution given by Equation (1) of this section.

4. A die is thrown four times. Find the probability that two threes, a four, and a six are obtained.

13.3 Distribution of Objects into Boxes

Case I *Distribution of Similar Objects in Distinguishable Boxes*

EXAMPLE 1   In how many ways can three similar balls be placed in two different boxes?

Solution   Consider the product

$$(1 + ax + a^2x^2 + a^3x^3 + \cdots)(1 + bx + b^2x^2 + b^3x^3 + \cdots)$$

$$= 1 + (a + b)x + (a^2 + ab + b^2)x^2 + (a^3 + a^2b + ab^2 + b^3)x^3 + \cdots.$$

The terms in the coefficient of x^3 represents the different distributions of three similar balls into boxes A and B; that is, a^3 represents "all three balls in A," a^2b represents "two balls in A, one in B," etc. Clearly if we put $a = b = 1$, we get the *number* of such distributions. Thus we seek the coefficient of x^3 in $(1 + x + x^2 + x^3 + \cdots)^2$, that is, the coefficient of x^3 in $[(1 - x)^{-1}]^2$. This will be the coefficient of x^3 in

$$(1 - x)^{-2} = \sum_{r=0}^{\infty} \binom{2 + r - 1}{r} x^r,$$

that is, $\binom{2 + 3 - 1}{3} = \binom{4}{3} = 4$.

EXAMPLE 2   In how many ways can two similar balls be placed in three different boxes?

Solution   Consider the product

$$(1 + ax + a^2x^2 + \cdots)(1 + bx + b^2x^2 + \cdots)(1 + cx + c^2x^2 + \cdots)$$

$$= 1 + (a + b + c)x + (a^2 + b^2 + c^2 + ab + bc + ca)x^2 + \cdots.$$

The terms in the coefficient of x^2 show the various possible distributions of two balls in three boxes. Their number is thus found to be (setting $a=b=c=1$) the coefficient 6 of x^2 in the expansion of $(1 + x + x^2 + \cdots)^3$ or $(1 - x)^{-3}$.

These examples suggest the following theorem.

THEOREM 1   The number of ways of placing r similar objects into n distinguishable boxes is $\binom{n+r-1}{r}$.

Proof   As in the examples, the required answer is the coefficient of x^r in the expansion of $(1 + x + x^2 + \cdots)^n = (1 - x)^{-n}$.
Since

$$(1 - x)^{-n} = \sum_{r=0}^{\infty} \binom{n+r-1}{r} x^r,$$

the required coefficient is $\binom{n+r-1}{r}$.

The problem solved in the above theorem has been discussed twice before in disguised form in Sections 4.3 and 6.4. It is easy to see that the number of ways of placing r similar objects into n distinguishable boxes is the number of solutions of the linear equation

$$x_1 + x_2 + \cdots + x_n = r$$

in nonnegative integers x_i. From Section 4.1 we know that the answer is $\binom{n+r-1}{n-1} = \binom{n+r-1}{r}$.
The number $\binom{n+r-1}{r}$ is also the number of lists of length r which can be made from n objects (repetitions allowed). If we consider the objects to be boxes and enter a box in the list for every object that is put in the box we see the one-to-one correspondence between distributions and lists.
The methods of this section have wider applicability. For example, they can often be applied to problems such as those considered in Section 4.2 in which the number of available objects is limited.

EXAMPLE 3   In how many ways can we distribute six similar balls among boxes A, B, C if A gets at most three balls, B at least two, and C gets at least one but no more than three.

Solution   We may translate the problem into one of linear equations with solutions bounded above and below and solve it with the methods of Section 4.2. Or the answer is the coefficient of x^6 in

$$(1 + x + x^2 + x^3)(x^2 + x^3 + x^4 + x^5 + x^6)(x + x^2 + x^3),$$

that is, the coefficient of x^3 in

$$(1 + x + x^2 + x^3)(1 + x + x^2 + x^3)(1 + x + x^2).$$

This coefficient is $3 + 3 + 2 + 1 = 9$.

The reader is referred to Niven's book[1] for a discussion of the following cases.

Case II   *Distribution of different objects in distinguishable boxes.*

Case III   *Distribution of different objects in indistinguishable boxes.*

Case IV   *Distribution of mixed objects in unlike boxes.*

E X E R C I S E 13.3

1.   Show why the number of ways of placing three similar balls into four boxes is the coefficient of x^3 in $(1 - x)^{-4}$.

2.   In how many ways can 20 similar balls be placed in five boxes?

3.   In how may ways can the balls be placed in Problem 2 if

    (a)   each box is to contain at least one ball,
    (b)   each box is to contain at least two balls?

4.   In how many ways can twelve apples be given to six teachers if each receives not more than three apples nor fewer than one apple?

5.   Solve Example 3 using the methods of Section 4.2.

6.*   If $a_1, a_2, \ldots, a_k$ are distinct prime numbers, how many different products can be formed by taking j of them at a time? (Products may include repeated factors.)

7.*   Use the methods of this section to solve Problem 11 of Exercise 4.2.

13.4 Stirling Numbers

The factorial

$$\lambda^{(n)} = \lambda(\lambda - 1) \cdots (\lambda - n + 1)$$

---

[1] I. Niven, "Mathematics of Choice," Random House, New York, 1965.

gives rise to a polynomial in λ of degree n when multiplied out. The coefficient of λ^r is denoted by $s(n, r)$, so that

$$\lambda^{(n)} = \sum_{r=0}^{n} s(n, r)\,\lambda^r; \tag{1}$$

$s(n, r)$ is called a **Stirling number of the first kind.**

Clearly $s(n, n) = 1$ and $s(n, 0) = 0$ for all n. Also $s(n, r) = 0$ for $r > n$ and $r < 0$. Now

$$\lambda^{(n)} = \lambda^n + s(n, n - 1)\lambda^{n-1} + \cdots + s(n, 2)\lambda^2 + s(n, 1)\lambda$$
$$\lambda(\lambda^{(n)}) = \lambda^{n+1} + s(n, n - 1)\lambda^n + s(n, n - 2)\lambda^{n-1} + \cdots + s(n, 1)\lambda^2$$
$$n\lambda^{(n)} = n\lambda^n + n\,s(n, n - 1)\lambda^{n-1} + \cdots + n\,s(n, 2)\lambda^2 + n\,s(n, 1)\lambda.$$

Thus, subtracting corresponding members of the last two equations,

$$\lambda^{(n)}(\lambda - n) = \lambda^{n+1} + \sum_{r=0}^{n} [s(n, r - 1) - n\,s(n, r)]\lambda^r,$$

$$\text{(taking } s(n, -1) = 0),$$

$$= \sum_{r=0}^{n+1} [s(n, r - 1) - n\,s(n, r)]\lambda^r,$$

$$\text{(taking } s(n, n + 1) = 0).$$

But

$$\lambda^{(n+1)} = [\lambda(\lambda - 1)\cdots(\lambda - n - 1)](\lambda - n) = \lambda^{(n)}(\lambda - n),$$

and

$$\lambda^{(n+1)} = \sum_{r=0}^{n+1} s(n + 1, r)\lambda^r.$$

Consequently, the Stirling numbers obey the recurrence relation

$$s(n + 1, r) = s(n, r - 1) - n\,s(n, r). \tag{2}$$

Using the iteration method described in Section 5.2 we can build up a table of Stirling numbers as far as desired (Table 13.1). Such a table makes it

TABLE 13.1

n \ r	0	1	2	3	4	5	6
1	0	1					
2	0	-1	1				
3	0	2	-3	1			
4	0	-6	11	-6	1		
5	0	24	-50	35	-10	1	
6	0	-120	274	-225	85	-15	1

easy to write a sum of factorial functions as an ordinary polynomial. For example,

$$\lambda^{(5)} + 4\lambda^{(4)} + 3\lambda^{(3)} = \lambda^5 + [-10 + 4(1)]\lambda^4 + [35 + 4(-6) + 3(1)]\lambda^3$$
$$+ [-50 + 4(11) + 3(-3)]\lambda^2 + [24 + 4(-6) + 3(2)]\lambda$$
$$= \lambda^5 - 6\lambda^4 + 14\lambda^3 - 15\lambda^2 + 6\lambda.$$

EXERCISE 13.4

1. (a) Extend the table of Stirling numbers to $n = 10$.
   (b)^c Write a computer program to evaluate Stirling numbers by means of the recurrence relation (2) of this section.
   (c)^c Extend the table of Stirling numbers to $n = 25$.

2.* Polynomials in λ can be expressed in terms of $\lambda^{(1)}, \lambda^{(2)}, \lambda^{(3)}, \ldots$ provided $\lambda, \lambda^2, \lambda^3, \ldots$ have been expressed in this form. Hence, let

$$\lambda^n = \sum_{r=0}^{n} t(n, r)\lambda^{(r)}. \tag{3}$$

   (i) Derive a recurrence relation for $t(n, r)$ analogous to (2).
   (ii) Construct a table of values of $t(n, r)$.
   (iii) Use the table constructed in (ii) to express the polynomial $\lambda^4 - 4\lambda^3 + 6\lambda^2 - 3\lambda$ (see page 19) in terms of factorials.

13.5 Gaussian Binomial Coefficients[2]

The **Gaussian binomial coefficient** $\begin{bmatrix} n \\ r \end{bmatrix}$ is defined by

$$\begin{bmatrix} n \\ r \end{bmatrix} = \frac{q^n - 1}{q - 1} \cdot \frac{q^{n-1} - 1}{q^2 - 1} \cdot \frac{q^{n-2} - 1}{q^3 - 1} \cdots \frac{q^{n+1-r} - 1}{q^r - 1}, \qquad (0 < r \le n)$$

$$\begin{bmatrix} n \\ 0 \end{bmatrix} = 1.$$

In the definition, q is a variable or formal mark.

EXAMPLE 1

(a) $\begin{bmatrix} 3 \\ 1 \end{bmatrix} = \dfrac{q^3 - 1}{q - 1} = 1 + q + q^2.$

---

[2] The work of this section was derived from a lecture presented at Waterloo in 1970 by Professor G. Polya of Stanford University.

(b) $\begin{bmatrix} 3 \\ 2 \end{bmatrix} = \dfrac{q^3 - 1}{q - 1} \cdot \dfrac{q^2 - 1}{q^2 - 1} = \dfrac{q^3 - 1}{q - 1} = 1 + q + q^2.$

(c) $\begin{bmatrix} 3 \\ 3 \end{bmatrix} = \dfrac{q^3 - 1}{q - 1} \cdot \dfrac{q^2 - 1}{q^2 - 1} \cdot \dfrac{q - 1}{q^3 - 1} = 1.$

The results in this section illustrate the remarkable analogy between the Gaussian binomial coefficients and the ordinary binomial coefficients. Theorem 1 shows a direct connection between the two types of binomial coefficients.

THEOREM 1   For the Gaussian binomial coefficient, $\begin{bmatrix} n \\ r \end{bmatrix}$,

$$\lim_{q \to 1} \begin{bmatrix} n \\ r \end{bmatrix} = \binom{n}{r}.$$

Proof Apply L'hôpital's rule.

$$\lim_{q \to 1} \begin{bmatrix} n \\ r \end{bmatrix} = \lim_{q \to 1} \frac{q^n - 1}{q - 1} \cdot \frac{q^{n-1} - 1}{q^2 - 1} \cdots \frac{q^{n-r+1} - 1}{q^r - 1}$$

$$= \lim_{q \to 1} \frac{nq^{n-1}}{1} \cdot \frac{(n-1)q^{n-2}}{2q} \cdots \frac{(n-r+1)q^{n-r}}{rq^{r-1}}$$

$$= \frac{n(n-1)\cdots(n-r+1)}{1 \cdot 2 \cdots r} = \binom{n}{r}.$$

THEOREM 2   The Gaussian binomial coefficients $\begin{bmatrix} n \\ r \end{bmatrix}$ obey the relation

$$\begin{bmatrix} n \\ r \end{bmatrix} = \begin{bmatrix} n \\ n - r \end{bmatrix}.$$

EXAMPLE 2   Show that (a) $\begin{bmatrix} 6 \\ 4 \end{bmatrix} = \begin{bmatrix} 6 \\ 2 \end{bmatrix}$ and (b) $\begin{bmatrix} 7 \\ 3 \end{bmatrix} = \begin{bmatrix} 7 \\ 4 \end{bmatrix}$.

Solution

(a) $\begin{bmatrix} 6 \\ 4 \end{bmatrix} = \dfrac{q^6 - 1}{q - 1} \cdot \dfrac{q^5 - 1}{q^2 - 1} \cdot \dfrac{q^4 - 1}{q^3 - 1} \cdot \dfrac{q^3 - 1}{q^4 - 1}$

$\qquad = \dfrac{q^6 - 1}{q - 1} \cdot \dfrac{q^5 - 1}{q^2 - 1} = \begin{bmatrix} 6 \\ 2 \end{bmatrix}.$

(b) $\begin{bmatrix} 7 \\ 4 \end{bmatrix} = \dfrac{q^7 - 1}{q - 1} \cdot \dfrac{q^6 - 1}{q^2 - 1} \cdot \dfrac{q^5 - 1}{q^3 - 1} \cdot \dfrac{q^4 - 1}{q^4 - 1}$

$\qquad = \dfrac{q^7 - 1}{q - 1} \cdot \dfrac{q^6 - 1}{q^2 - 1} \cdot \dfrac{q^5 - 1}{q^3 - 1} = \begin{bmatrix} 7 \\ 3 \end{bmatrix}.$

The following theorem can be used to define Gaussian binomial coefficients. Then the definition stated at the beginning of this section becomes a theorem.

THEOREM 3   For the Gaussian binomial coefficient $\begin{bmatrix} n \\ r \end{bmatrix}$,

$$(1 + x)(1 + qx)(1 + q^2x) \cdots (1 + q^{n-1}x) = \sum_{r=0}^{n} \begin{bmatrix} n \\ r \end{bmatrix} q^{r(r-1)/2} x^r.$$

EXAMPLE 3   Show that

$$(1 + x)(1 + qx)(1 + q^2x) = \sum_{r=0}^{3} \begin{bmatrix} 3 \\ r \end{bmatrix} q^{r(r-1)/2} x^r.$$

Solution

$$(1 + x)(1 + qx)(1 + q^2x)$$
$$= 1 + (1 + q + q^2)x + (q + q^2 + q^3)x^2 + (1 \cdot q \cdot q^2)x^3$$

$$= 1 + \frac{q^3 - 1}{q - 1}x + q \cdot \frac{q^3 - 1}{q - 1}x^2 + q^3x^3$$

$$= \begin{bmatrix} 3 \\ 0 \end{bmatrix} + \begin{bmatrix} 3 \\ 1 \end{bmatrix}x + \begin{bmatrix} 3 \\ 2 \end{bmatrix}qx^2 + \begin{bmatrix} 3 \\ 3 \end{bmatrix}q^3x^3$$

$$= \sum_{r=0}^{3} \begin{bmatrix} 3 \\ r \end{bmatrix} q^{r(r-1)/2} x^r$$

since $\frac{1}{2}0(0 - 1) = 0$, $\frac{1}{2}(1)(0) = 0$, $\frac{1}{2}(2)(1) = 1$, and $\frac{1}{2}(3)(2) = 3$.

The following theorem may be verified by substitution.

THEOREM 4   For the Gaussian binomial coefficient $\begin{bmatrix} n \\ r \end{bmatrix}$,

$$\begin{bmatrix} n \\ r \end{bmatrix} + \begin{bmatrix} n \\ r - 1 \end{bmatrix} q^{n+1-r} = \begin{bmatrix} n + 1 \\ r \end{bmatrix}.$$

EXAMPLE 4   Show that $\begin{bmatrix} 5 \\ 2 \end{bmatrix} + \begin{bmatrix} 5 \\ 1 \end{bmatrix}q^4 = \begin{bmatrix} 6 \\ 2 \end{bmatrix}$.

Solution

$$\begin{bmatrix} 5 \\ 2 \end{bmatrix} + \begin{bmatrix} 5 \\ 1 \end{bmatrix}q^4 = \frac{q^5 - 1}{q - 1}\left[\frac{q^4 - 1}{q^2 - 1} + q^4\right]$$

$$= \frac{q^5 - 1}{q - 1} \cdot \frac{q^6 - 1}{q^2 - 1}$$

$$= \frac{q^6 - 1}{q - 1} \cdot \frac{q^5 - 1}{q^2 - 1} = \begin{bmatrix} 6 \\ 2 \end{bmatrix}.$$

The reader should note that if we take the limit, as q approaches 1, in Theorems 2, 3, and 4, we obtain the results

(i) $$\binom{n}{r} = \binom{n}{n-r},$$

(ii) $$(1 + x)^n = \sum_{r=0}^{n} \binom{n}{r} x^r,$$

(iii) $$\binom{n}{r} + \binom{n}{r-1} = \binom{n+1}{r},$$

respectively. These results were proved independently for binomial coefficients.

THEOREM 5 The Gaussian binomial coefficient $\begin{bmatrix} n \\ r \end{bmatrix}$ can be expressed in the form

$$\begin{bmatrix} n \\ r \end{bmatrix} = A_0 + A_1 q + A_2 q^2 + \cdots + A_{r(n-r)} q^{r(n-r)}$$

where the A_i are integers and $A_0 + A_1 + \cdots + A_{r(n-r)} = \binom{n}{r}$.

Note that

$$\begin{bmatrix} n \\ 1 \end{bmatrix} = \frac{q^n - 1}{q - 1} = 1 + q + q^2 + \cdots + q^{n-1}$$

and the coefficients sum to $n = \binom{n}{1}$.

EXAMPLE 5

$$\begin{bmatrix} 6 \\ 2 \end{bmatrix} = \begin{bmatrix} 6 \\ 4 \end{bmatrix} = \frac{q^6 - 1}{q - 1} \cdot \frac{q^5 - 1}{q^2 - 1}$$

$$= \frac{q^6 - 1}{q^2 - 1} \cdot \frac{q^5 - 1}{q - 1}$$

$$= (q^4 + q^2 + 1)(q^4 + q^3 + q^2 + q + 1)$$

$$= 1 + q + 2q^2 + 2q^3 + 3q^4 + 2q^5 + 2q^6 + q^7 + q^8.$$

The sum of the coefficients is

$$1 + 1 + 2 + 2 + 3 + 2 + 2 + 1 + 1 = 15 = \binom{6}{2}.$$

Recall that in Section 2.7 we showed a connection between the binomial coefficients $\binom{n}{r}$ and the number of paths from the origin to the point $(n-r, r)$. We conclude this section by pointing out the connection between the coefficients A_i of q^i in the expansion of $\begin{bmatrix} n \\ r \end{bmatrix}$ and the *areas* under the paths from the origin to the point $(n - r, r)$.

EXAMPLE 6   Show that among the $\binom{3}{1} = 3$ different paths from the origin to the point (2, 1), one has zero area under it, one has area one unit under it, and one has area two units under it. Note that

$$\begin{bmatrix} 3 \\ 1 \end{bmatrix} = \frac{q^3 - 1}{q - 1} = 1 + q + q^2.$$

Solution   The paths are shown in Figure 13.1.

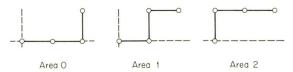

Area 0   Area 1   Area 2

Figure 13.1

EXAMPLE 7   Show that the coefficient of q^i in $\begin{bmatrix} 4 \\ 2 \end{bmatrix}$ is the number of paths from the origin to the point (2, 2) which have areas of i units under them (subtend areas of i units).

Solution

$$\begin{bmatrix} 4 \\ 2 \end{bmatrix} = \frac{q^4 - 1}{q - 1} \cdot \frac{q^3 - 1}{q^2 - 1} = (q^2 + 1)(q^2 + q + 1)$$

$$= q^4 + q^3 + 2q^2 + q + 1.$$

There are $\binom{4}{2} = 6$ distinct path from the origin to (2, 2). These are shown in Figure 13.2. Now compare the number of paths subtending an area of i units with the coefficient of q^i.

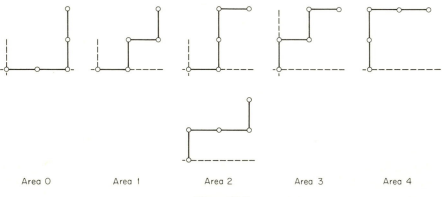

Area 0   Area 1   Area 2   Area 3   Area 4

Figure 13.2

THEOREM 6   A particle can move one unit at a time in either the positive x or positive y direction and goes from the origin to the point $(n - r, r)$. The number of paths it can take which subtend an area of i units is A_i where A_i is the coefficient of q^i in the expansion of the Gaussian binomial coefficient $\begin{bmatrix} n \\ r \end{bmatrix}$,

$$\begin{bmatrix} n \\ r \end{bmatrix} = A_0 + A_1 q + \cdots + A_i q^i + \cdots + A_{r(n-r)} q^{r(n-r)}.$$

We present the above theorem without proof. However we make the following comments.

(i) $\sum_i A_i = \binom{n}{r}$ = number of distinct paths from the origin to the point $(n - r, r)$ (Section 2.7).

(ii)   The possible areas subtended, in square units, range over the integral values from 0 to $r(n - r)$, corresponding to the powers of q in the expansion of $\begin{bmatrix} n \\ r \end{bmatrix}$.

(iii)   If we label each move of one unit to the right in such a path with a plus sign and each move of one unit upwards with a minus sign, then each path is represented by a string of $n - r$ plus signs and r minus signs.

(iv)   In each such string of plus and minus signs, every minus sign is followed by a certain number of plus signs. Every time a minus sign is followed by a plus sign, one unit of area is generated because the particle has moved first upward and then to the right.

(v)   The number t of times a minus sign is followed by a plus sign in such a string of plus and minus signs corresponds to the power q^t in the expansion of $\begin{bmatrix} n \\ r \end{bmatrix}$.

EXAMPLE 8   Illustrate the comments (i)–(v) above using the distinct paths from the origin to the point (4, 2).

Solution   The number of such paths is $\binom{6}{2} = 15$. These are shown in Figure 13.3 along with the corresponding arrays of plus and minus signs and the corresponding terms of q^i. Finally,

$$\begin{bmatrix} 6 \\ 2 \end{bmatrix} = \frac{q^6 - 1}{q - 1} \cdot \frac{q^5 - 1}{q^2 - 1}$$

$$= 1 + q + 2q^2 + 2q^3 + 3q^4 + 2q^5 + 2q^6 + q^7 + q^8.$$

The reader should check the number of times a minus sign is followed by a plus sign in each array. For example, in the string $- + - + + +$ corresponding to q^7, the first minus sign is followed by a total of four plus signs and the second minus sign by three for a total of seven.

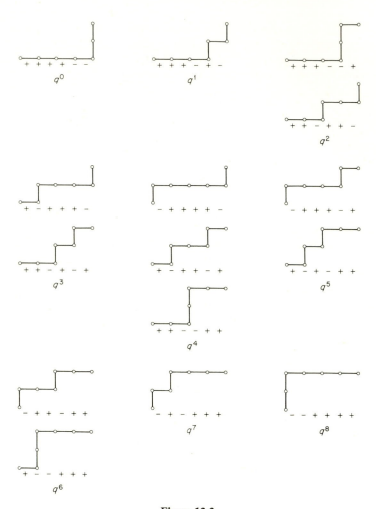

Figure 13.3

EXERCISE 13.5

1. Write expressions for

(a) $\begin{bmatrix} 3 \\ 1 \end{bmatrix}$ (b) $\begin{bmatrix} 3 \\ 2 \end{bmatrix}$ (c) $\begin{bmatrix} 4 \\ 3 \end{bmatrix}$ (d) $\begin{bmatrix} 7 \\ 4 \end{bmatrix}$ (e) $\begin{bmatrix} 10 \\ 8 \end{bmatrix}$.

2. Verify that

(a) $\begin{bmatrix} 6 \\ 5 \end{bmatrix} = \begin{bmatrix} 6 \\ 1 \end{bmatrix}$ (b) $\begin{bmatrix} 7 \\ 3 \end{bmatrix} = \begin{bmatrix} 7 \\ 4 \end{bmatrix}$.

3. Verify that

(a) $\lim_{q \to 1} \begin{bmatrix} 3 \\ 2 \end{bmatrix} = \begin{pmatrix} 3 \\ 2 \end{pmatrix}$ (b) $\lim_{q \to 1} \begin{bmatrix} 6 \\ 4 \end{bmatrix} = \begin{pmatrix} 6 \\ 4 \end{pmatrix}$.

4. Prove Theorem 2 of this section.

5. Verify Theorem 3 of this section for (a) $n = 4$, (b) $n = 5$.

6. Verify that

(a) $\begin{bmatrix} 6 \\ 2 \end{bmatrix} + \begin{bmatrix} 6 \\ 1 \end{bmatrix} q^5 = \begin{bmatrix} 7 \\ 2 \end{bmatrix}$ (b) $\begin{bmatrix} 4 \\ 3 \end{bmatrix} + \begin{bmatrix} 4 \\ 2 \end{bmatrix} q^2 = \begin{bmatrix} 5 \\ 3 \end{bmatrix}$.

7. Prove Theorem 4 of this section.

8. Express the Gaussian binomial coefficients in problem (1) as power series in q.

9. Verify Theorem 6 of this section for

(a) $\begin{bmatrix} 5 \\ 2 \end{bmatrix}$ (b) $\begin{bmatrix} 6 \\ 3 \end{bmatrix}$ (c) $\begin{bmatrix} 7 \\ 2 \end{bmatrix}$.

Show the different paths from the origin to the appropriate point in each case, label each path with a string of plus and minus signs, and show the correspondence with the terms q^i in $\begin{bmatrix} n \\ r \end{bmatrix}$.

More Generating Functions and Difference Equations

Perhaps the most useful tool in enumeration is the use of generating functions in conjunction with difference equations. This chapter contains important examples which further illustrate the techniques involved. The first section applies generating functions and difference equations to the study of partitions of integers and the second uses them to determine the number of ways of triangulating a polygon. Section 14.3 uses these techniques in some further work on random walks which were introduced in Chapter 1. Finally, Section 14.4 discusses a general class of difference equations. Special cases of this class of equations play an important role in the mathematics of operations research. In this section we mention waiting lines, birth or death processes, and renewal theory.

14.1 The Partition of Integers

A **partition** of a positive integer n is a representation of n as a sum of positive integers. The order of the integers in the sum is immaterial. In this section we show how partitions are related to generating functions and difference equations.

EXAMPLE 1   Show that the number of partitions of 4 is the coefficient of x^4 in the expansion of

$$\phi(x) = (1 + x + x^2 + x^3 + x^4)(1 + x^2 + x^4)(1 + x^3)(1 + x^4).$$

Solution   There are five partitions of 4, namely,

$$1 + 1 + 1 + 1, \quad 1 + 1 + 2, \quad 2 + 2, \quad 1 + 3, \quad 4.$$

Each term in x^4 in the expansion of $\phi(x)$ is made up of a number of x's from the first factor (0, 1, 2, 3, or 4, and these correspond to 1's in the partition of 4), a number of x^2's from the second factor (0, 1, or 2, and these correspond to 2's in the partition of 4), a number of x^3's from the third factor (0 or 1; these correspond to 3's in the partition), and, finally, a number of x^4's from the last factor (0 or 1, and these correspond to 4's in the partition of 4). Thus, for example, the term x^4 which is the product of x^4 from the first factor and 1's from the remaining three factors corresponds to the partition $1 + 1 + 1 + 1$ since we are choosing four x's from the first factor. The term x^4 which is the product of x from the first factor, $x^0 = 1$ from the second, x^3 from the third, and $x^0 = 1$ from the fourth corresponds to the partition $1 + 0 + 3 + 0 = 1 + 3$.

This example leads us to the following theorem which we present without proof.

THEOREM 1   If $a, b, \ldots, h$ are different positive integers, then the coefficient of x^n in the expansion of

$$(1 + x^a + x^{2a} + \cdots)(1 + x^b + x^{2b} + \cdots) \cdots (1 + x^h + x^{2h} + \cdots)$$

(each bracket includes all factors up to x^n) is the number of partitions of n with summands $a, b, \ldots, h$.

EXAMPLE 2   Illustrate Theorem 1 with $n = 9$ and using the summand 3.

Solution   By Theorem 1 the number of partitions of 9 with summands 3 is the coefficient of x^9 in

$$(1 + x^3 + x^6 + x^9).$$

This coefficient is 1 corresponding to the partition $3 + 3 + 3$.

Note that in this example, the coefficient of x^{10} is 0 and this corresponds to the number of partitions of 10 with summands 3.

EXAMPLE 3   Illustrate Theorem 1 for $n \leq 10$ using summands 2 and 3.

Solution

$$(1 + x^2 + x^4 + x^6 + x^8 + x^{10})(1 + x^3 + x^6 + x^9)$$
$$= 1 + x^2 + x^3 + x^4 + x^5 + 2x^6 + x^7 + 2x^8 + 2x^9 + 2x^{10} + \cdots.$$

The coefficients of x^i, $i \le 10$ in the above expansion correspond to partitions using summands 2 and 3:

$$
\begin{array}{ll}
10 = 2 + 2 + 3 + 3 & 5 = 3 + 2 \\
 = 2 + 2 + 2 + 2 + 2 & \\
9 = 3 + 3 + 3 & 4 = 2 + 2 \\
 = 2 + 2 + 2 + 3 & \\
8 = 2 + 3 + 3 & 3 = 3 \\
 = 2 + 2 + 2 + 2 & \\
7 = 2 + 2 + 3 & 2 = 2 \\
6 = 3 + 3 & \\
 = 2 + 2 + 2 & 1 \text{ (no partition).}
\end{array}
$$

Note that the partition $2 + 3 + 3$ of 8 corresponds to the term in x^8 in the above expansion which is the product of x^2 from the first factor and $x^6 = (x^3)^2$ from the second.

EXAMPLE 4   What product can be used to give the number of partitions of 24 with summands 3, 5, or 7?

Solution

$$(1 + x^3 + x^6 + \cdots + x^{24})(1 + x^5 + \cdots + x^{20})(1 + x^7 + x^{14} + x^{21}).$$

EXAMPLE 5   Find the number of ways that a 50-cent piece can be changed into pennies, nickels, dimes, and quarters.

Solution   Rephrase the problem: "Find the number of partitions of 50 with summands 1, 5, 10, or 25." The required number is the coefficient of x^{50} in the expansion of

$$(1 + x + x^2 + \cdots + x^{50})(1 + x^5 + x^{10} + \cdots + x^{50})$$
$$\times (1 + x^{10} + x^{20} + \cdots + x^{50})(1 + x^{25} + x^{50}).$$

We use the symbol $p(n)$ to denote the number of partitions of n. From Example 1, $p(4) = 5$.

EXERCISE 14.1

1.   Find the terms of the product

$$(1 + x^2 + x^4)(1 + x^3 + x^6)(1 + x^4 + x^8)$$

and interpret the result in connection with partitions of the integers 1, 2, 3, 4 with summands 2, 3, and 4.

2. Write the terms of the product

    $$(1 + x + x^2 + \cdots + x^{12})(1 + x^2 + x^4 + \cdots + x^{12})$$
    $$\times (1 + x^3 + x^6 + x^9 + x^{12})$$

    up to x^{12} and interpret the result to give the partitions of the integers 1, ..., 12 with summands 1, 2, and 3.

3. Does the coefficient of x^8 in Problem 1 give all the partitions of the integer 8 with summands 2, 3, and 4? Explain.

4. Show that the coefficient of x^8 in Problem 2 is the number of solutions, in nonnegative integers, of

    $$x + 2y + 3z = 8.$$

5. Show that, if a and b are distinct positive integers, the coefficient of x^n in the expansion of

    $$(1 + x^a + x^{2a} + \cdots)(1 + x^b + x^{2b} + \cdots)$$

    is the number of solutions, in nonnegative integers, of

    $$xa + yb = n.$$

6. Suggest a method for finding the number of ways a dollar bill can be changed into (a) pennies, nickels, and dimes; (b) pennies, nickels, dimes and quarters; (c) pennies, nickels, dimes, quarters, and half-dollars.

7. Show how to use the result of Problem 6(b) to find the number of ways $1.03 can be made up of pennies, nickels, dimes, and quarters.

8. Describe how to find the number of solutions, in nonnegative integers, of

    (a) $x + 5y + 10z + 25t = 83$
    (b) $2x + 3y + 5z + 7t = 35.$

9. Evaluate $p(1)$, $p(2)$, $p(3)$, $p(5)$, $p(7)$, $p(8)$.

10.^{c} Write a computer program for finding partitions of an integer n employing Theorem 1. (This requires a subroutine for multiplying polynomials.)

11.^{c} Check your program in Problem 10^c by finding the partitions in Example 3.

12.^{c} Use the computer to obtain numerical solutions to Problems 6, 7, and 8.

14.2 Triangulation of Convex Polygons

This section illustrates the use of relatively complicated recursion relations.

THEOREM 1   Let C_n be the number of ways of dividing a convex $(n + 2)$-gon into triangles by drawing nonintersecting diagonals. Define $C_0 = 1$. Then

$$C_{n+1} = \sum_{k=0}^{n} C_k C_{n-k}.$$

Proof   Consider an $(n + 3)$-gon and single out one side. A polygon with one side singled out or identified is a **rooted polygon**. C_{n+1} is the number of ways of dividing an $(n + 3)$-gon into triangles as described above. Joining the ends of the rooted side of the $(n + 3)$-gon with one of the other $n + 1$ vertices produces a triangle (which we will refer to as a basic triangle) and two other component figures, A and B, as in Figure 14.1.

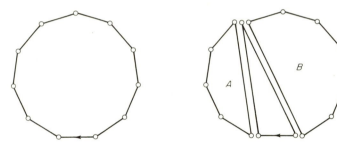

Figure 14.1

Let component A have $k + 2$ sides; then B has $n - k + 2$ sides. The two diagonals drawn are duplicated in the base triangle and the two components so there is a total of $n + 7$ edges in the three components.

The number of ways to triangulate A is C_k and to triangulate B is C_{n-k}. Any triangulation of A together with any triangulation of B together with the base triangle provides a triangulation of the original $(n + 3)$-gon. Hence there are $C_k C_{n-k}$ ways to triangulate the $(n + 3)$-gon with the particular base triangle as the partition between the components.

Since every contribution to C_{n+1} contains one triangle on the rooted side and there are $n + 1$ possible partitioning base triangles, the various possibilities correspond to the values of $k + 2$ (the number of sides of A) from 2 to $n + 2$; that is, k ranges from 0 to n and

$$C_{n+1} = \sum_{k=0}^{n} C_k C_{n-k}.$$

This proves the theorem.

We may compute $C_0 = 1$, $C_1 = 1$, $C_2 = 2$, $C_3 = 5$, etc. Theorem 2 develops a formula for C_n.

THEOREM 2   Let C_n be the number of ways of dividing a convex $(n + 2)$-gon into triangles by drawing nonintersecting diagonals and define $C_0 = 1$. Then

$$C_n = \frac{(2n)!}{n!(n + 1)!}.$$

Proof   Define

$$C(x) = \sum_{n=0}^{\infty} C_n x^n. \tag{1}$$

Now

$$C_{n+1} x^{n+1} = \sum_{k=0}^{n} C_k C_{n-k} x^{n+1}$$

so that

$$\sum_{n=0}^{\infty} C_{n+1} x^{n+1} = x \sum_{n=0}^{\infty} \sum_{k=0}^{n} C_k C_{n-k} x^n.$$

It follows that

$$C(x) - 1 = x[C(x)]^2$$

or

$$x[C(x)]^2 - C(x) + 1 = 0.$$

There are two solutions for $C(x)$, namely

$$C_1(x) = \frac{1 - \sqrt{1 - 4x}}{2x} \quad \text{and} \quad C_2(x) = \frac{1 + \sqrt{1 - 4x}}{2x}.$$

It can be verified using the binomial theorem that

$$C_1(x) = \sum_{n=0}^{\infty} \frac{(2n)!}{n!(n + 1)!} x^n$$

and

$$C_2(x) = \frac{1}{x} - C_1(x) = \frac{1}{x} - \sum_{n=0}^{\infty} \frac{(2n)!}{n!(n + 1)!} x^n.$$

Notice that $C_2(x)$ is not defined for $x = 0$. Since (1), which defines the C_n, is a power series, it follows that $C(x) = C_1(x)$ so that

$$C_n = \frac{(2n)!}{n!(n + 1)!},$$

as required.

EXERCISE 14.2*

1. When a convex n-gon is decomposed into triangles by diagonals which do not intersect, how many triangles result? Prove that this is always the same no matter how the "triangulation" is carried out. How many diagonals are drawn in such a procedure? Prove this number is also constant.

2. If $2n$ points are marked on the circumference of a circle, in how many different ways can these points be joined in pairs by n chords which do not intersect within the circle?

3. Addition is a "binary" operation. From a very strict viewpoint, brackets need to be inserted in a sum consisting of more than two terms in order to indicate the "binary" steps in the evaluation; for example, $[a + (b + c)] + d$. In the case of four addends, a, b, c, d, there exist five ways in which these brackets might be inserted without changing the order:

$$[a + (b + c)] + d, \quad [(a + b) + c] + d, \quad a + [b + (c + d)],$$
$$a + [(b + c) + d], \quad (a + b) + (c + d).$$

Determine the number of ways, $f(n)$, of imposing brackets on an expression containing n addends without changing their order.

4. Determine a formula for $f(n)$, the number of regions into which the interior of a convex n-gon is divided by its diagonals, provided no three diagonals are concurrent.

5. How many convex r-gons can be drawn, all of whose vertices are vertices of a given convex n-gon and all of whose sides are diagonals of the n-gon?

6. If

$$a_k = \binom{k}{0} + \binom{k+1}{2} + \cdots + \binom{k+j}{2j} + \cdots + \binom{2k}{2k}, \quad k \geq 0,$$

$$b_k = \binom{k}{1} + \binom{k+1}{3} + \cdots + \binom{k+j}{1+2j} + \cdots + \binom{2k-1}{2k-1}, \quad k > 0,$$

show that

$$a_k + b_k = b_{k+1}$$

and

$$a_k + b_{k+1} = a_{k+1}.$$

7. If $a_n(x) = (1 - x)^{n+1} \sum_{k=0}^{\infty} k^n x^k$, show that

$$a_n(x) = nxa_{n-1}(x) + x(1 - x)\frac{d[a_{n-1}(x)]}{dx},$$

and verify that

$$a_0(x) = 1, \qquad a_2(x) = x + x^2,$$
$$a_1(x) = x, \qquad a_3(x) = x + 4x^2 + x^3.$$

8.^{c} Write a computer program to evaluate recursively the integers C_n defined in Theorem 1.

9.^{c} Use the computer program of Problem 8^c to find the number of ways of dividing (a) a hexagon, (b) a 10-gon into triangles by drawing non-intersecting diagonals.

14.3 Random Walks

Random walks were first introduced in Chapter 1. In this section we consider other examples of random walks and show how they are related to the binomial theorem.

EXAMPLE 1   Assume that a particle begins at the origin of a cartesian coordinate system and can move a distance of one unit at a time in either the positive x or positive y direction. Suppose that the probability of moving one unit in the positive x direction is p and of moving one unit in the positive y direction is q, where $p + q = 1$. Let $P^k(m, n)$ (with $m + n = k$) denote the probability that the particle is at the point (m, n) at the end of k moves. Then we have the following recurrence relation:

$$P^k(m, n) = pP^{k-1}(m - 1, n) + qP^{k-1}(m, n - 1). \tag{1}$$

The probabilities may be calculated successively. The first few, evaluated from first principles, are

$P^0(0, 0) = 1$	$P^1(1, 0) = p$	$P^2(2, 0) = p^2$	$P^3(3, 0) = p^3$
	$P^1(0, 1) = q$	$P^2(1, 1) = 2pq$	$P^3(2, 1) = 3p^2q$
		$P^2(0, 2) = q^2$	$P^3(1, 2) = 3pq^2$
			$P^3(0, 3) = q^3$.

Note that $P^3(2, 1)$ is the coefficient of x^2y in the expansion of $(px + qy)^3$ and similarly for the other terms. This suggests the following theorem.

THEOREM 1   A particle moving in a random walk begins at the origin and moves either one unit in the positive x direction (with probability p) or one unit in the positive y direction (with probability $q = 1 - p$) in each interval of time. Let $P^k(m, n)$, $m + n = k$, denote the probability that the particle is at the point (m, n) at the end of k intervals of time. Then

$$P^k(m, n) = \binom{k}{m} p^m q^n.$$

Proof Let

$$F_k(x, y) = \sum_{m, n = 0}^{\infty} P^k(m, n) x^m y^n, \qquad \text{with} \quad m + n = k,$$

$$= p \sum_{m, n = 0}^{\infty} P^{k-1}(m - 1, n) x^m y^n$$

$$+ q \sum_{m, n = 0}^{\infty} P^{k-1}(m, n - 1) x^m y^n, \qquad \text{[from (1)]}$$

$$= px \sum_{\substack{n = 0 \\ m - 1 = 0}}^{\infty} P^{k-1}(m - 1, n) x^{m-1} y^n$$

$$+ qy \sum_{\substack{m = 0 \\ n - 1 = 0}}^{\infty} P^{k-1}(m, n - 1) x^m y^{n-1}$$

$$= (px + qy) F_{k-1}(x, y)$$
$$\vdots$$
$$= (px + qy)^{k-1} F_1(x, y)$$
$$= (px + qy)^k F_0(x, y)$$
$$= (px + qy)^k$$

since $F_0(x, y) = P^0(0, 0) + 0 = 1$.
It follows that $P^k(m, n) = \binom{k}{m} p^m q^n$ as required.

EXAMPLE 2   A particle begins at the origin and each interval of time moves either one unit in the positive x direction, with probability $\frac{1}{6}$, or one unit in the positive y direction, with probability $\frac{5}{6}$. What is the probability that the particle (a) passes through $(4, 1)$; (b) does not pass through $(4, 1)$; (c) passes through both $(4, 1)$ and $(5, 0)$; (d) passes through both $(4, 1)$ and $(2, 1)$; (e) passes through neither $(4, 1)$ nor $(2, 1)$; (f) passes through one of $(4, 1)$ or $(2, 1)$ but not both.

Solution

(a) $P^5(4, 1)$ = coefficient of x^4y in $(\frac{1}{6}x + \frac{5}{6}y)^2$

$$= \binom{5}{4}\left(\frac{1}{6}\right)^4\left(\frac{5}{6}\right) = \frac{25}{6^5}.$$

(b) $1 - \dfrac{25}{6^5}$

(c)   If the particle passes through $(4, 1)$ then it cannot pass through $(5, 0)$ so that the required probability is zero.

(d) $P^3(2, 1) = \binom{3}{2}\left(\frac{1}{6}\right)^2\left(\frac{5}{6}\right) = \dfrac{15}{6^3}.$

To reach $(4, 1)$ from $(2, 1)$ requires two consecutive moves in the positive x direction. The probability of this is $(\frac{1}{6})^2$. Hence the probability the particle passes through both points is

$$\frac{15}{6^3}\cdot\frac{1}{6^2} = \frac{15}{6^5}.$$

(e)   Let A = "particle passes through $(4, 1)$,"
      B = "particle passes through $(2, 1)$."
Then $\Pr(\alpha\beta)$ = probability the particle passes through neither point

$$= 1 - \Pr(A) - \Pr(B) + \Pr(AB)$$

$$= 1 - \frac{25}{6^5} - \frac{15}{6^3} + \frac{15}{6^5} \text{ from (a) and (d).}$$

(f) $\alpha B + A\beta = (U - A)B + A(U - B)$
$$= B - AB + A - AB$$
$$= A + B - 2AB$$

$$\Pr(\alpha B + A\beta) = \frac{25}{6^5} + \frac{15}{6^3} - \frac{30}{6^5} = \frac{535}{6^5}.$$

Theorem 1 tells us that $P^k(m, n)$ is the coefficient of x^my^n in the expansion of $(px + qy)^k$. The movement of the particle as described in Example 1 and Theorem 1 is called a **binomial random walk** because of its connection with the binomial coefficients. The generating function $\phi(x, y) = (px + qy)^k$ gives us the probabilities $P^k(m, n)$.

EXAMPLE 3   Suppose a particle begins at the origin and moves one unit at a time along the x axis. Let p be the probability that it moves one unit in the positive direction and q the probability that it moves one unit in the negative direction in any interval of time.

Let $P^k(n)$ denote the probability that the position of the particle is $(n, 0)$ at the end of k intervals of time. A few values of this probability distribution for points near the origin are given below:

$$
\begin{array}{llll}
P^0(0) = 1 & P^1(1) \;\; = p & P^2(2) \;\; = p^2 & P^3(3) \;\; = p^3 \\
& P^1(-1) = q & P^2(0) \;\; = 2pq & P^3(1) \;\; = 3p^2q \\
& & P^2(-2) = q^2 & P^3(-1) = 3pq^2 \\
& & & P^3(-3) = q^3.
\end{array}
$$

Note that $P^3(-1)$ is the coefficient of x^{-1} in the expansion of

$$(px + qx^{-1})^3.$$

This suggests the following theorem.

THEOREM 2   A particle moving in a random walk begins at the origin and moves either one unit in the positive x direction, with probability p, or one unit in the negative x direction, with probability $q = 1 - p$, in each interval of time. Let $P^k(n)$ denote the probability that the particle is at the point $(n, 0)$ at the end of k intervals of time. Then

$$P^k(n) = \text{coefficient of } x^n \text{ in the expansion of } (px + qx^{-1})^k.$$

Proof   The probabilities $P^k(n)$ obey the recurrence relation

$$P^{k+1}(n) = pP^k(n - 1) + qP^k(n + 1). \tag{2}$$

Let

$$F_k(x) = \sum_{n=-\infty}^{\infty} P^k(n)x^n.$$

Then, using the recurrence relation (2),

$$F_k(x) = \sum_{n=-\infty}^{\infty} pP^{k-1}(n - 1)x^n + \sum_{n=-\infty}^{\infty} qP^{k-1}(n + 1)x^n, \qquad \text{[from (2)]}$$

$$= px \sum_{n-1=-\infty}^{\infty} P^{k-1}(n - 1)x^{n-1} + qx^{-1} \sum_{n+1=-\infty}^{\infty} P^{k-1}(n + 1)x^{n+1}$$

$$= (px + qx^{-1})F_{k-1}(x)$$

$$\vdots$$

$$= (px + qx^{-1})^{k-1}F_1(x)$$

$$= (px + qx^{-1})^k F_0(x)$$

$$= (px + qx^{-1})^k$$

since $F_0(x) = P^0(0)x^0 + 0 = 1$.

Hence $P^k(n) = $ coefficient of x^n in $(px + qx^{-1})^k$, as stated in the theorem.

EXAMPLE 4   Find the probability that the particle in Theorem 2 is at $(1, 0)$ after four moves if $p = \frac{1}{3}$ and $q = \frac{2}{3}$.

Solution

$$P^4(1) = \text{coefficient of } x \text{ in } \left(\frac{1}{3}x + \frac{2}{3}x^{-1}\right)^4 \text{ or } \frac{x^4}{3^4}(1 + 2x^{-2})^4$$

$$= \frac{1}{3^4} \times \text{coefficient of } x^{-3} \text{ in } (1 + 2x^{-2})^4$$

$$= 0.$$

EXAMPLE 5   In Example 4, find the probability that the particle is to the right of the origin after four moves.

Solution   After four moves the possible locations of the particle are $(4, 0), (2, 0), (0, 0), (-2, 0),$ and $(-4, 0)$. Since the only two locations to the right of the origin are $(4, 0)$ and $(2, 0)$, we need only calculate $P^4(4)$ and $P^4(2)$.

$$P^4(4) = \text{coefficient of } x^4 \text{ in } \left(\frac{1}{3}x + \frac{2}{3}x^{-1}\right)^4$$

$$= \frac{1}{3^4} \times \text{constant term in } (1 + 2x^{-2})^4$$

$$= \frac{1}{3^4}.$$

$$P^4(2) = \frac{1}{3^4} \times \text{coefficient of } x^{-2} \text{ in } (1 + 2x^{-2})^4 = 1 + 4(2x^{-2}) + \cdots$$

$$= \frac{1}{3^4} \times 8.$$

Hence the required probability is $9/3^4 = 1/9$.

The following theorems, presented without proof, give some further variations of the simple random walk.[1]

THEOREM 3   A particle moving in a random walk begins at the origin and each interval of time moves one unit in the positive x direction with probability p, moves one unit in the negative x direction with probability q,

---

[1] For a further discussion of Theorems 3 and 4 see G. Berman, A class of difference equations, *SIAM Review*, **7** (1965), 513–525.

or remains where it is with probability d, where $p + q + d = 1$. Let $Q^k(n)$ denote the probability that the particle is at the point $(n, 0)$ at the end of k intervals of time. Then $Q^k(n) =$ coefficient of x^n in the expansion of

$$(px + d + qx^{-1})^k.$$

THEOREM 4   A particle moving in a random walk begins at the origin and moves either two units in the positive x direction (with probability p) or one unit in the negative x direction (with probability $q = 1 - p$) in each interval of time. Let $R^k(n)$ denote the probability that the particle is at the point $(n, 0)$ at the end of k intervals of time. Then

$$R^k(n) = \text{coefficient of } x^n \text{ in the expansion of } (px^2 + qx^{-1})^k.$$

EXERCISE 14.3

1. A particle moving in a random walk begins at the origin and moves one unit at a time along the x axis. The probability that it moves in the positive direction is $\frac{1}{4}$ and in the negative direction $\frac{3}{4}$. What is the probability that the particle

    (a)   is at $(2, 0)$ at the end of four moves;
    (b)   is at $(-2, 0)$ at the end of five moves;
    (c)   is on the negative side of the origin at the end of five moves;
    (d)   does not enter the negative side of the axis at all during the first five moves?

2. A particle moving in a random walk begins at the origin and moves one unit at a time along the x axis. The probability that it moves in the positive direction is $\frac{1}{2}$, in the negative direction $\frac{1}{3}$, and the probability that it remains stationary during an interval of time is $\frac{1}{6}$. Find the probability that the particle

    (a)   is at $(3, 0)$ at the end of five intervals of time;
    (b)   is at $(-2, 0)$ at the end of five intervals of time;
    (c)   is on the negative side of the origin at the end of five intervals of time;
    (d)   does not enter the negative side of the axis at all during the first five intervals of time.

3. A particle moving in a random walk begins at the origin and moves along the x axis. The probability of a one-unit step in the positive direction is $\frac{1}{10}$, of a two-unit step in the positive direction is $\frac{1}{5}$, of a three-unit step in

the negative direction is $\frac{3}{10}$, and of a delay is $\frac{2}{5}$ in each interval of time. What is the probability that at the end of five intervals of time the particle

(a)   is at the origin;
(b)   is at $(-1, 0)$?

4.   A particle moving in a random walk begins at the origin and moves one unit at a time in either the positive x or positive y direction each interval of time with probabilities $\frac{3}{5}$ and $\frac{2}{5}$ respectively. What is the probability that the particle

(a)   passes through the point $(3, 4)$;
(b)   fails to pass through $(5, 2)$;
(c)   passes through both $(1, 2)$ and $(4, 3)$;
(d)   passes through one of $(1, 2)$, $(4, 3)$ but not the other;
(e)   fails to pass through either $(1, 2)$ or $(4, 3)$;
(f)   never leaves the x axis?

5.* Prove Theorem 3 of this section.

6.* Prove Theorem 4 of this section.

14.4 A Class of Difference Equations[2]

The difference equations (1) and (2) of Section 14.3 are special cases of the class of difference equations which satisfy the recurrence relation

$$f_{k+1}(\lambda) = \sum_{i,j} a_{ij} f_k(\lambda - b_{ij} e_i). \tag{1}$$

The f_k in (1) are real-valued functions defined on the integral points $\lambda = (\lambda_1, \lambda_2, \ldots, \lambda_n)$ of Euclidean n-space E_n (the λ_i are integers); the a_{ij} are real numbers, and the b_{ij} are integers, not necessarily positive. The symbol e_i represents the point

$$(0, 0, \ldots, 1, \ldots, 0)$$

with 1 in the ith position and 0's elsewhere, and $\lambda - b_{ij} e_i$ denotes the point (in vector notation)

$$(\lambda_1, \lambda_2, \ldots, \lambda_{i-1}, \lambda_i - b_{ij}, \lambda_{i+1}, \ldots, \lambda_n).$$

---

[2] For a further discussion of the work of this section see G. Berman, A class of difference equations, *SIAM Review*, **7** (1965), 513–525.

The summation is taken over all integral values of j ranging from 1 to n_i (the n_i are given integers) and over all integral values of i ranging from 1 to n.

The following examples illustrate the difference equations (1) for special cases. In particular, Example 1 includes a generalization of Pascal's formula.

EXAMPLE 1   Suppose k objects are chosen from n types of objects, λ_i being chosen of type i for $i = 1, \ldots, n$, so that $\sum \lambda_i = k$, and we wish to find the number of possible distinguishable choices, $f_k(\lambda)$.

Every choice λ of $k + 1$ objects can be built up from a choice $\lambda - e_i$ of k objects by adjoining an element of type i. Further, no choices constructed in this way are the same. Thus

$$f_{k+1}(\lambda) = \sum_i f_k(\lambda - e_i), \qquad (2)$$

and the boundary conditions are $f_0(0) = 1; f_0(\lambda) = 0, \lambda \neq 0$. It can be verified that the multinomial coefficients

$$f_k(\lambda) = \binom{k}{\lambda_1, \lambda_2, \ldots, \lambda_n} = \frac{k!}{\lambda_1! \lambda_2! \cdots \lambda_n!}$$

satisfy equation (2).

EXAMPLE 2   Let f_k denote the number of subsets of a set S_k containing k distinguishable objects. If another element a is added to S_k, the enlarged set S_{k+1} contains $k + 1$ elements, exactly one-half of whose subsets contain a. Thus

$$f_{k+1} = 2f_k.$$

Since the number of subsets of the empty set is $1, f_0 = 1$, and it is immediate that $f_k = 2^k$.

Notice that this is the particular case of Equation (1) in which $n = 0$, that is, the corresponding space E_0 is a single point even though Equation (1) was formulated for $n > 0$.

EXAMPLE 3   (Waiting Lines)   Suppose n types of objects are arriving at a depot or warehouse in a random manner such that in any given fixed time-interval the probability that j objects of type i arrive is p_{ij}, $i = 1, 2, \ldots, n$, $j = 0, 1, 2, \ldots, n_i$, $\sum_{i, j} p_{ij} = 1$. Let $f_k(\lambda)$ denote the probability that at the end of the kth time period λ_i objects of type i have arrived, where $\lambda = (\lambda_1, \lambda_2, \ldots, \lambda_n)$.

At the end of $k + 1$ intervals the number of arrivals will be described by λ (that is, λ_i objects of type i) if at the end of the kth interval the number of arrivals is described by $\lambda - je_i$ and there were j arrivals of type i during the $(k + 1)$st interval. It follows that

$$f_{k+1}(\lambda) = \sum_{i,j} p_{ij} f_k(\lambda - je_i). \tag{3}$$

In this case arbitrary initial conditions can be assumed.

The following examples illustrate two interpretations of Equation (3).

EXAMPLE 4   (Birth or Death Processes)   Multiple-type birth or death processes, for example of male or female pigs, can be interpreted as non-homogeneous waiting lines. Although the process is continuous, it can be approximated at discrete times. The parameters p_{ij} can be interpreted as the probability of j births (or deaths) of type i occurring during one time interval, and $f_k(\lambda)$ as the probability that the number of births (or deaths) occurring during k periods is described by λ.

EXAMPLE 5   (Renewal Theory)   Consider a large nonhomogeneous population containing n types of objects with a finite lifetime, for example, light bulbs or tubes of various types. Suppose replacements are made at regular intervals, say at times $t_0, t_1, t_2, \ldots$, and the probability that j objects of type i will require replacement at time t_k is p_{ij}, independent of k. Let $f_k(\lambda)$ denote the probability that λ_i objects of type i require replacement up to and including time t_k (or in any k consecutive intervals).

EXERCISE 14.4

1. Suppose three objects are chosen from five different types of objects. Find the number of possible distinguishable choices $f_k(\lambda)$, where $\lambda = (\lambda_1, \lambda_2, \lambda_3, \lambda_4, \lambda_5)$.

2. Verify that Pascal's formula is a special case of Equation (1) of this section.

3. If five objects are chosen from three types of objects, in how many ways can there be three of the first type and one each of the remaining two types?

4.* Construct special cases of Equation (3) of this section for $n = 1$ and 2 and evaluate $f_0(\lambda)$, $f_1(\lambda)$, and $f_2(\lambda)$ iteratively using the boundary conditions $f_0(0) = 1$, $f_0(\lambda) = 0$ for $\lambda \neq 0$.

Fibonacci Sequences

In this chapter we review some of the basic ideas connected with Fibonacci sequences and present a few illustrative examples. We show how Fibonacci sequences occur in Pascal's triangle and in sequences of plus and minus signs. Then we illustrate how Fibonacci sequences occur in applications by two examples, counting (fertile) hares and in finding maxima and minima of real-valued functions.

15.1 Representations of Fibonacci Sequences

In Section 5.1 we introduced the Fibonacci sequence

$$\{1, 1, 2, 3, 5, 8, 13, 21, \ldots\} \tag{1}$$

which is determined by the difference equation

$$F(n) = F(n-1) + F(n-2) \tag{2}$$

and the initial conditions

$$F(1) = F(2) = 1. \tag{3}$$

By choosing different initial conditions, in general,

$$F(1) = a, \qquad F(2) = b \tag{4}$$

we obtain different sequences which are also called Fibonacci sequences, each of which satisfies the difference equation (2).

EXAMPLE 1   Let $a = 2$ and $b = -1$ in Equations (4). Then applying (2) iteratively, we obtain the Fibonacci sequence

$$\{2, -1, 1, 0, 1, 1, 2, \ldots\}$$

which contains all the terms of the sequence (1). This is not always the case. If we choose $a = -1$ and $b = 3$ in equations (4), we obtain the Fibonacci sequence

$$\{-1, 3, 2, 5, 7, \ldots\}.$$

The general term of the sequence defined by Equations (2) and (4) can be expressed in terms of the particular sequence (1). This is done in Theorem 1.

THEOREM 1   Let $f_1, f_2, \ldots$ denote the terms of the Fibonacci sequence

$$\{1, 1, 2, 3, 5, 8, 13, \ldots\}$$

and let $F_1, F_2, \ldots$ denote the terms of the Fibonacci sequence defined by the equations

$$F(n) = F(n-1) + F(n-2), \qquad F(1) = a, \quad F(2) = b.$$

Then

$$F_n = f_{n-2} a + f_{n-1} b \qquad \text{for} \quad n \geq 3. \tag{5}$$

This theorem can be readily verified for small values of n and can be proved by induction.

It was shown in Section 5.1 that if $F(n)$ satisfies the difference equation (2) then

$$F(n) = c_1 \left(\frac{1 + \sqrt{5}}{2} \right)^n + c_2 \left(\frac{1 - \sqrt{5}}{2} \right)^n. \tag{6}$$

With the initial conditions (3) we found that $c_1 = -c_2 = 1/\sqrt{5}$, and this determines an explicit formula for f_n. If we use $F(1) = a$ and $F(2) = b$, we obtain the general formula which is stated in Theorem 2.

THEOREM 2   Let $F_1, F_2, \ldots$ denote the terms of the Fibonacci sequence determined by the equations

$$F(n) = F(n-1) + F(n-2), \qquad F(1) = a, \quad F(2) = b.$$

Then

$$F_n = \frac{a(\omega - 1) + b}{\sqrt{5}} \omega^{n-1} + \frac{a(1 - \bar{\omega}) - b}{\sqrt{5}} \bar{\omega}^{n-1} \tag{7}$$

where

$$\omega = \frac{1 + \sqrt{5}}{2} \quad \text{and} \quad \bar{\omega} = \frac{1 - \sqrt{5}}{2}.$$

Another method for obtaining the explicit formula for f_n employs an interesting recursive representation[1] of the Fibonacci sequence $f_1, f_2, \ldots$. Let

$$x^2 = x + 1. \tag{8}$$

Then it is easy to verify by induction that

$$x^n = f_n x + f_{n-1},$$

since, multiplying by x and using (8) yields

$$\begin{aligned}
x^{n+1} &= f_n x^2 + f_{n-1} x \\
&= f_n(x + 1) + f_{n-1} x \\
&= (f_n + f_{n-1})x + f_n \\
&= f_{n+1} x + f_n.
\end{aligned}$$

The formula for the terms of the sequence is now easy to deduce. The roots of Equation (8) are ω and $\bar{\omega}$ as defined above and we have

$$\omega^n = f_n \omega + f_{n-1}$$
$$\bar{\omega}^n = f_n \bar{\omega} + f_{n-1}$$

so that

$$f_n = \frac{\omega^n - \bar{\omega}^n}{\omega - \bar{\omega}}.$$

In Section 6.4, it was shown that the generating function f for the Fibonacci sequence (1) satisfies the equation

$$(1 - x - x^2)f(x) = F(1) + (F(2) - F(1))x,$$

with $F(1) = F(2) = 1$. By using the initial conditions (4) we obtain the generating function of the general Fibonacci sequence as stated in Theorem 3.

THEOREM 3   Let f denote the generating function of the Fibonacci sequence determined by the equations

$$F(n) = F(n - 1) + F(n - 2), \qquad F(1) = a, \quad F(2) = b.$$

[1] E. Just, A note on the nth term of the Fibonacci sequence, *Math. Magazine*, **44** (1971), 199.

Then

$$f(x) = \frac{a + (b - a)x}{1 - x - x^2}.$$

EXERCISE 15.1

1. Prove Theorem 1 of this section.

2. Verify formula (7) of this section in the following ways: (i) by induction; (ii) by substituting from Formula (6) with $c_1 = -c_2 = 1/\sqrt{5}$ for f_n into (5); (iii) by noting that c_1 and c_2 satisfy the equations

$$Ac_1 + Bc_2 = a$$
$$A^2c_1 + B^2c_2 = b$$

obtained from (6) with the aid of (4); (iv) by expressing (8) in partial fraction form and obtaining the coefficient of x^n.

3.[c] Write a computer program for evaluating F_n. See how many values you obtain before running into problems. Try using formulas (5) or (7) with various values of a and b. Would it be better to represent the integers as decimal fractions?

4. Prove that $F(n + 1) = 2F(n - 1) + F(n - 2)$ for any Fibonacci sequence.

15.2 Diagonal Sums of the Pascal Triangle

Draw diagonals in the Pascal triangle as shown in Figure 15.1. If we add up the numbers between the parallel lines, we note that the sums are precisely the Fibonacci numbers of sequence (1). These sums are shown at the left of the Pascal triangle.

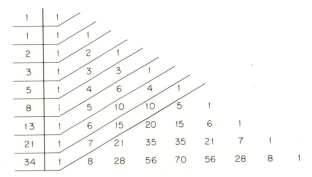

Figure 15.1

It is easy to see why this is so. For example, consider the diagonal whose sum is 34. Notice that $7 = 6 + 1$ with 6 and 1 in each of the two preceding diagonals. Similarly $15 = 10 + 5$ with 10 and 5 in each of the two preceding diagonals. Also $10 = 4 + 6$ and finally the two additional 1's in the diagonal correspond to 1's in each of the two preceding diagonals. Thus the sum of the numbers in this particular (ninth) diagonal is the sum of the numbers in the two preceding (seventh and eighth) diagonals.

It is left as an exercise to prove by induction that the above property of the Pascal triangle holds in general.

EXERCISE 15.2

1. Use induction to show that the sums of the diagonals of the Pascal triangle as indicated in Figure 15.1 are the Fibonacci numbers.

2. Show that the elements of the nth diagonal of the Pascal triangle are

$$\binom{n-1}{0}, \binom{n-2}{1}, \binom{n-3}{2}, \dots, \binom{n-k-1}{k}, \dots.$$

3. Use the representation of Problem 2 to show that the sums of the diagonals of the Pascal triangle are the Fibonacci numbers.

15.3 Sequences of Plus and Minus Signs

In this section we consider sequence of plus and minus signs. By a PM sequence we shall mean a sequence of plus or minus signs with no two minus signs adjacent.

EXAMPLE 1

(i) There are two PM sequences of one plus or minus sign, namely $\{+\}$ and $\{-\}$.
(ii) There are three PM sequences of two plus or minus signs, namely $\{+, +\}, \{+, -\}$, and $\{-, +\}$.
(iii) There are five PM sequences with $n = 3$, namely, $\{+, +, +\}, \{+, +, -\}$, $\{+, -, +\}, \{-, +, +\}, \{-, +, -\}$.

Let F_n denote the number of different PM sequences of n plus or minus signs. From Example 1 we see that $F_1 = 2$, $F_2 = 3$, and $F_3 = 5$. These are the first three terms of the Fibonacci sequence defined by the equations

$$F(n) = F(n-1) + F(n-2), \qquad F(1) = 2, \quad F(2) = 3.$$

In this section we shall prove that F_n is, in fact, the nth term of this sequence of PM sequences. We do this by a sequence of simple theorems.

THEOREM 1   The number of PM sequences containing k plus signs and r minus signs is $C(k + 1, r)$.

We demonstrate the proof of this theorem in the following example.

EXAMPLE 2   Find the number of PM sequences of six plus and three minus signs.

Solution   Write the six plus signs between t's as follows:

$$t + t + t + t + t + t + t.$$

Replacing three of the t's by minus signs (then dropping the remaining t's) yields a desired combination. This can be done in $C(7, 3)$ ways.

THEOREM 2   The number F_n of PM sequences with n terms is given by

$$F_n = C(n + 1, 0) + C(n, 1) + C(n - 1, 2) + \cdots.$$

The proof follows immediately from Theorem 1 by noting that the number of minus signs can be 0, 1, 2, ... This argument is demonstrated in the following example.

EXAMPLE 3   Find the number of PM sequences of seven plus or minus signs.

Solution   Consider the following array of seven t's:

$$t \; t \; t \; t \; t \; t \; t$$

If each t can be replaced by either a plus or a minus sign, we would have 2^7 possible arrays. However we are interested only in those arrays in which minus signs are not adjacent. We have the following possibilities:

seven plus signs;
six plus signs, one minus sign;
five plus, two minus signs;
four plus, three minus signs;
three plus, four minus signs.

Thus by Theorem 1,

$$F_7 = C(8, 0) + C(7, 1) + C(6, 2) + C(5, 3) + C(4, 4)$$
$$= 1 + 7 + 15 + 10 + 1 = 34.$$

THEOREM 3   If F_n is the number of PM sequences with n terms then

$$F_n = F_{n-1} + F_{n-2}, \qquad F_1 = 2, \quad F_2 = 3,$$

and hence $\{F_n\}$ is a Fibonacci sequence.

Proof   Every sequence of plus or minus signs with no two minus signs adjacent must begin with either $+$ or $-$.

(i)   An initial $+$ can be followed by any of the F_{n-1} sequence of $n-1$ plus or minus signs with no two minus signs adjacent.

(ii)   An initial $-$ must be followed by a $+$ and then the $-+$ can be followed by any of the F_{n-2} sequences of $n-2$ plus or minus signs with no two minus signs adjacent.

It follows that $F_n = F_{n-1} + F_{n-2}$ and we have seen in Example 1 that $F_1 = 2$ and $F_2 = 3$.

EXERCISE 15.3

In the following problems assume the men are indistinguishable and that the women are indistinguishable but that it is possible to tell a man from a woman.

1.   In how many ways can 12 men and 7 women sit in a row without having two women in adjacent chairs?

2.   If the people in Problem 1 seat themselves at random, what is the probability that at least two women sit in adjacent chairs?

3.   In how many ways can 19 people sit in a row without having two men in adjacent chairs?

4.   In how many ways can 19 red or blue balls be placed in a row without having two red balls adjacent?

5.   Ten people get on an elevator at the ground floor. The elevator then stops at floors 1 to 23 inclusive. If no more than one person gets off at each stop, in how many ways can the people leave the elevator so that no two people get off at adjacent floors?

6.   In how many eight digit numbers, formed from the single digits (to be used only once)

$$1, 2, 3, 4, 5, 6, 7, 9$$

will there not be at least two successive even digits?

7. In how many of the permutations of the letters of the word *Lackawanna*, taken all at a time, will there not be two a's adjacent?

8. Prove Theorem 1 of this section (a) by induction, (b) by the method suggested in Example 2.

15.4 Counting Hares[2]

Fibonacci first discovered his famous sequence about 1202 in studying the problem of breeding rabbits. He assumed that rabbits take one month to mature and another month to bring forth a litter. For simplicity, he assumed that exactly one couple from each litter, along with their parents, were to be retained for future generations. He also assumed that any rabbits which died or lost their fertility were to be replaced.

Let F_n denote the number of pairs alive at the end of n months if we start with one pair of newborn rabbits. In examining the data one might conclude that the number of rabbits would double every two months. However this is not the case.

Since we started with one pair of rabbits, there would be no increase during the first two months, that is, $F_0 = F_1 = 1$. At the end of the second month there would be an additional pair obtained from a litter so that $F_2 = 2$. At the end of the third month the original pair would produce an additional newborn pair, but the second pair would not. Hence $F_3 = 3$. At the end of the fourth month the pair produced during the third month would not produce a litter, but the two remaining pairs would, that is, two new pairs would be added and $F_5 = 5$.

The general situation is clear. At the end of the nth month only the F_{n-2} pairs of rabbits alive before the $(n-1)$st month would produce a litter, so that at the end of the nth month, there would be the F_{n-1} pairs of rabbits alive at the beginning of the month plus the pairs from the new litters, that is,

$$F_n = F_{n-1} + F_{n-2}.$$

EXERCISE 15.4

1. Suppose we started with three pairs of rabbits and made the same assumptions as Fibonacci. How many pairs of rabbits would there be at the end of seven months?

[2] H. S. M. Coxeter, The golden section, phyllotaxis and Wythoff's Game, *Scripta Mathematica*, **20** (1954), 135–143.

2.   Find the number of pairs of rabbits G_n at the end of n months if we started with k pairs of newborn rabbits.

3.   Repeat Problem (2) assuming that h of the pairs of rabbits are one year old at the beginning.

4.* Assume that two pairs of rabbits are kept from each litter. Let H_n denote the number of pairs in existence after n months (leaving the other assumptions of Fibonacci the same). Find (i) the difference equation satisfied by H_n, (ii) the corresponding generating function, (iii) a formula for H_n.

5.* Repeat Problem 4 with the assumption that p pairs of rabbits are kept from each litter.

15.5 Maximum or Minimum of a Unimodal Function

In this section we show how Fibonacci numbers are related to the problem of finding the maximum value or the minimum value of a unimodal function in an efficient manner without employing calculus. This method also enables us to locate the minimum of functions such as $|x|$ which cannot be handled by calculus.

It is sufficient to restrict our attention to finding the minimum value of a function since we can find the maximum of f by finding the minimum of $-f$.

Everyone who has studied calculus understands the meaning of a local minimum. The function f has a local minimum at x^* if $f(x^*) \leq f(x)$ for all x in some interval about x^*. In Figure 15.2 the points A, B, C, and D are local minima of f defined between a and b. Note that calculus *cannot* be used to find

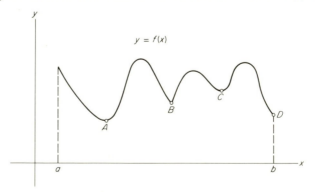

Figure 15.2

the local minima at B and D since the function does not have a derivative at B and the value of the derivative at D is not 0.

A function f (defined on some interval $a \le x \le b$) is **unimodal** if it has exactly one local minimum or maximum in this region. Examples of unimodal functions are shown in Figure 15.3. The minima of these functions cannot be found using calculus.

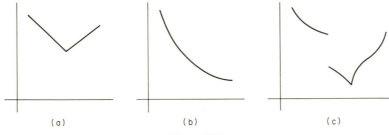

(a)   (b)   (c)

Figure 15.3

Suppose we know that the function f defined on an interval $a \le x \le b$ is unimodal. Consider the values of f at x_1 and x_2 with $x_1 < x_2$ in the given interval. There are three possibilities (Figure 15.4) according as $f(x_1) < f(x_2)$, $f(x_1) = f(x_2)$, or $f(x_1) > f(x_2)$. If $f(x_1) < f(x_2)$, as in Figure 15.4a, then the minimum of f must occur in the interval $a \le x \le x_2$. If $f(x_1) = f(x_2)$, as in Figure 15.4b, the minimum must occur in the interval $x_1 \le x \le x_2$. Finally, if $f(x_1) > f(x_2)$, as in Figure 15.4c, the minimum must occur in the interval $x_1 \le x \le b$.

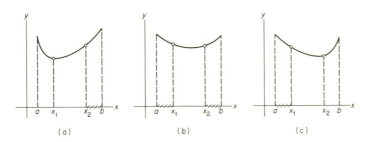

(a)   (b)   (c)

Figure 15.4

The interval in which the minimum is known to lie is called the **interval of uncertainty**. In the beginning the interval of uncertainty is $a \le x \le b$. When the value $f(x_1)$ is known, the interval of uncertainty is still $a \le x \le b$. However, once $f(x_2)$ is calculated for a second point x_2 of the interval, the interval of uncertainty is reduced to one of the intervals

$$a \le x < x_1, \qquad x_1 \le x \le x_2, \qquad \text{or} \qquad x_2 < x \le b.$$

In practical application the second interval occurs so infrequently that it can be included in either of the other cases say the first. If this is done, the remaining interval already has $f(x)$ evaluated at one point. Thus by evaluating f at an additional point, the interval of uncertainty will be reduced. If an additional point is chosen in the resulting interval of uncertainty, the resulting interval of uncertainty will be still smaller.

In practice, we would continue until we obtain a given accuracy, for example, say, until we have the interval of uncertainty reduced to a width of 0.001 units. (The midpoint is 0.0005 units from each end.) Then if the last value of x chosen in the interval is 3.452, it is clearly the minimum to three decimals (or to four significant figures).

In 1953 Kiefer[3] showed how the sequence $\{x_1, x_2, \ldots\}$ of points should be selected so that at each stage the interval of uncertainty is the smallest possible independent of the particular unimodal function f under consideration. This method is referred to as the **Fibonacci algorithm for the minimum**.

We first note that if we desire a given accuracy of computation (the exact value can never be obtained by this method), we can restrict the points from which we choose the x_i to a discrete set. Thus we partition the interval $a \leq x \leq b$ into N equal intervals of width $\delta = (b - a)/N$. This yields the $N + 1$ points $\xi_0 = a, \xi_1 = a + \delta, \xi_2 = a + 2\delta, \ldots, \xi_N = b$.

EXAMPLE 1   If $a = 1$, $b = 5$, and $N = 8$, then $\delta = 0.5$ and the points of the subdivision are

$$1.0, 1.5, 2.0, 2.5, 3.0, 3.5, 4.0, 4.5, 5.0.$$

Suppose that N is a Fibonacci number F_n for some Fibonacci sequence and select the points

$$x_1 = a + \delta F_{n-2}, \qquad x_2 = a + \delta F_{n-1}$$

as shown in Figure 15.5. Since $F_n = F_{n-1} + F_{n-2}$, we have $b - x_2 = F_{n-2}\delta$

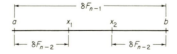

Figure 15.5

so that the points x_1, x_2 are symmetrically spaced relative to a and b. It follows either with $f(x_1) \leq f(x_2)$ or $f(x_1) > f(x_2)$ that the new interval of uncertainty has width δF_{n-1} (the original interval of uncertainty had width δF_n). The two

[3] J. Kiefer, Sequential minimax search for a maximum, *Proc. Am. Math. Soc.* **4** (1953), 503–506.

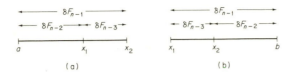

Figure 15.6

situations are shown in Figure 15.6. In either case,

$$x_2 - x_1 = \delta F_{n-1} - \delta F_{n-2} = \delta F_{n-3}.$$

Now select x_3 in Figure 15.6a at the point $a + \delta F_{n-3}$ and in Figure 15.6b at the point $x_1 + \delta F_{n-2} = b - \delta F_{n-3}$. We will then have the same situation as shown in Figure 15.5 but with $n - 1$ instead of n. Now either $f(x_1) \le f(x_3)$ or $f(x_1) > f(x_3)$ corresponding to Figure 15.6a or $f(x_2) \le f(x_3)$ or $f(x_2) > f(x_3)$ corresponding to Figure 15.6b and in any case we will reduce the interval of uncertainty to a new interval of width δF_{n-2}.

We continue in this way reducing the intervals of uncertainty to widths $\delta F_{n-3}, \delta F_{n-4}, \ldots$, until we reach an interval of width 2δ which is our final interval of uncertainty, as shown in Figure 15.7. Here the points x_j, $\bar{x}$,

Figure 15.7

and x_k are particular values of the x_i's. The point $\bar{x}$ is an approximation to x^*, the minimum point, and $f(\bar{x})$ is an approximation to $f(x^*)$, the minimum value of the function. Further, we know that $|\bar{x} - x^*| \le \delta$. Note that because of the symmetry of the construction, the positioning of x_1 determines all the remaining positions of the x_i for $i = 2, 3, \ldots, n + 1$. At each stage we have a point x_i in the interval and simply choose the symmetric point as the next point x_{i+1}, and continue in this way until the width of the interval of uncertainty is 2δ.

The following example will illustrate the Fibonacci search for a minimum which is described above. We choose a simple function for which we already know the required minimum. In this way we hope that the reader will be convinced of the validity and the ease of applying this method.

EXAMPLE 2   Find the minimum value of the function defined by $y = |x + 3|$ in the interval $(-4, 1)$.

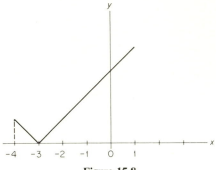

Figure 15.8

Solution   The sketch of the function in Figure 15.8 indicates that the required minimum value is 0 at $x = -3$. The width of the given interval $(-4, 1)$ is 5. Let us take $n = 10$ so that $F_{10} = 55$. We divide the interval $(-4, 1)$ into 55 intervals of width $\frac{1}{11}$ as indicated in Figure 15.9. Starting at -4, each

Figure 15.9

point of division can be labeled 0, 1, 2, ..., 54, 55 as shown, in terms of intervals of width $\frac{1}{11}$. Note that $b = 1$ and $F_{10}\delta = 55(\frac{1}{11})$. First select $x_1 = -4 + F_8\delta = -4 + 21(\frac{1}{11}) = -\frac{23}{11}$ and $x_2 = -4 + F_9\delta = -4 + 34(\frac{1}{11}) = -\frac{10}{11}$. These points are determined by marking off 21 units and 34 units of $\frac{1}{11}$ to the right of -4 (Figure 15.10). Now $f(x_1) = f(-\frac{23}{11}) = \frac{10}{11}$ and $f(x_2) = f(-\frac{10}{11}) = \frac{23}{11}$ so that $f(x_1) < f(x_2)$. This means that the minimum is in the interval $(-4, x_2)$ and we can discard the interval $(x_2, 1)$. We next select the point $x_3 = -4 + F_7\delta = -4 + 13(\frac{1}{11}) = -\frac{31}{11}$. This point is obtained graphically (Figure 15.10) by marking off $F_7 = 13$ units of $\frac{1}{11}$ to the right of -4. Again, $f(x_3) = \frac{2}{11}$ and $f(x_3) < f(x_1)$ so that we may discard the interval (x_1, x_2).

We continue in Figure 15.11 with the scale doubled. We show -4, x_3, and x_1. We select $x_4 = -4 + F_6\delta = -4 + 8(\frac{1}{11}) = -\frac{36}{11}$, sketching x_4 by marking

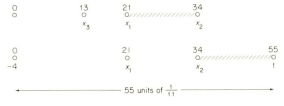

Figure 15.10

```
                    10  11  12  13
                     O   O  O///O
                    x₇  x₆  x₈  x₃

         8          10  11        13
         O///////O   O   O         O
         x₄         x₇  x₆        x₃

         8              11    13          16
         O               O    O//////////O
         x₄             x₆    x₃          x₅

         8                   13    16                21
         O                    O    O////////////////O
         x₄                  x₃    x₅                x₁

O                        8         13                21
 O////////////////////////O         O                 O
-4                       x₄         x₃                x₁
```

Figure 15.11

off $F_6 = 8$ units of $\frac{1}{11}$ to the right of -4. Since $f(x_4) = \frac{3}{11}, f(x_4) > f(x_3)$ so we discard the interval $(-4, x_4)$ and the remaining interval of uncertainty (x_4, x_1) has width 13δ. Now we select $x_5 = x_4 + F_6\delta$ or $x_5 = x_1 - F_5\delta$ and obtain in either case $x_5 = -\frac{28}{11}$. Here $f(x_5) = \frac{5}{11}$ and $f(x_3) < f(x_5)$ so we may discard the interval (x_5, x_1) and retain the interval of uncertainty (x_4, x_5) of width 8δ. Select $x_6 = x_4 + F_4\delta = -\frac{36}{11} + 3(\frac{1}{11}) = -\frac{33}{11}; f(x_6) = 0$ and $f(x_6) < f(x_3)$ so that we discard the interval (x_3, x_5) leaving an interval of uncertainty (x_4, x_3) of width 5δ. Select $x_7 = x_4 + F_3\delta = x_3 - F_4\delta = -\frac{34}{11}; f(x_7) = \frac{1}{11}$ and $f(x_7) > f(x_6)$ so that we discard the interval (x_4, x_7) and retain the interval of uncertainty (x_7, x_3) of width 3δ. Finally, $x_8 = x_7 + F_3\delta = -\frac{32}{11}; f(x_8) = \frac{1}{11}$ and $f(x_8) > f(x_6)$ so that we discard the interval (x_8, x_3), leaving the interval of uncertainty (x_7, x_8) of width 2δ. We may now state that the minimum value of the function defined by $y = |x + 3|$ in the interval $(-4, 1)$ is located approximately at $x = x_6 = -3$ with a possible maximum error of $\delta = \frac{1}{11}$, and the minimum value is approximately 0. In this particular example the minimum value happens to be exactly 0 at $x = -3$ but there is no way of knowing this from the algorithm. In general, we will obtain only an approximation to the location of the minimum with an interval of uncertainty of width δ.

We summarize the above procedure in an algorithm which can be readily computerized.

FIBONACCI ALGORITHM   (for minimizing a unimodal function).
Let f be defined in the interval $I_1 = \{x \mid a \le x \le b\}$ and suppose we wish to find the minimum point with an accuracy ε. Let $\delta < \varepsilon$ be such that

$$F_n\delta = b - a. \tag{1}$$

(The smallest value of n satisfying (1) is sufficient.) Let $x_1 = a + F_{n-2}\delta$.

Select x_2, the point symmetric to x_1 in I, and compare $f(x_1)$ and $f(x_2)$. If $f(x_1) \le f(x_2)$, let $I_2 = \{x | a \le x \le x_2\}$ and if $f(x_1) > f(x_2)$, let $I_2 = \{x | x_1 \le x \le b\}$. The interval I_2 will then contain either x_1 or x_2 as an interior point. Let x_3 denote the point symmetric to this interior point in I_2 and determine I_3 as described above. Continue in this way for $n - 2$ steps to obtain an interval of length $F_3 \delta = 2\delta$ with the point $\bar{x}$ interior.

This point $\bar{x}$ is an approximation to the minimum point x^* of the function with accuracy ε, that is, $|\bar{x} - x^*| < \varepsilon$, and $f(\bar{x})$ is an approximation to the minimum value of f in the given interval.

EXERCISE 15.5

1. Test the algorithm of this section by hand computation to find the minimum values of the following functions.

(a) $f(x) = |x|, -1 \le x \le 3$;
(b) $f(x) = |x|, 2 \le x \le 3$;
(c) $f(x) = |x - 4|, 0 \le x \le 6$;
(d) $f(x) = |x| + |x - 4|, 2 \le x \le 5$;
(e) $f(x) = 4 - x, 0 \le x < 2,$
   $= x + 1, 2 \le x \le 4$;
(f) $f(x) = 4 - x, 0 \le x < 2,$
   $= x^2, 2 \le x \le 3$.

2.^{c} Write a computer program for the Fibonacci algorithm. (Note that finding a subroutine to find the Fibonacci numbers was suggested in Exercise 15.1.)

3.^{c} Check your program in Problem 2 by finding the minima of the functions defined in Problem 1 to three decimal places.

4.^{c} Find the minimum of $f(x) = \sin x$, in the interval $0.1 \le x \le 0.3$.

16

Arrangements

In the previous chapters we have been primarily concerned with enumeration problems and with systems such as finite geometries that have particular properties. Another type of combinatorial problem is concerned with the *arrangement* of objects rather than with their number or properties. In this chapter we will consider a few important examples of arrangements of objects and properties of these arrangements.

Section 16.1 discusses the problem of finding systems of distinct representatives of sets of objects. In the remaining sections we consider arrangements having various symmetries. Latin squares and the existence of orthogonal latin squares are studied in Section 16.2. One of the results mentioned in this section was discovered only in the last decade. The Kirkman Schoolgirl Problem discussed in Section 16.3 has important modern applications to balanced incomplete block designs which are arrangements considered in Section 16.4. The difference sets of Section 16.5 are connected with designs and also with the projective planes which were studied in Chapter 11. Finally, in Sections 16.6 and 16.7, we mention magic squares and a generalization of latin squares called *Room squares*.

16.1 Systems of Distinct Representatives

In this section we begin with general arrangements of elements into sets and we consider the problem of selecting one element from each set to act as a representative of the set. We illustrate this problem with an example.

EXAMPLE 1   Let $U = \{1, 2, 3, 4\}$ and consider the subsets $S_1 = \{1\}$, $S_2 = \{1, 2, 3\}$, and $S_3 = \{3, 4\}$. Then $(1, 1, 3)$ is a system of representatives of the sets S_1, S_2, and S_3 respectively and $(1, 2, 3)$ is a system of distinct representatives. Notice that in each case one element is chosen from each set, but if we wish distinct representatives, 1 cannot be selected as a representative of S_2. If we had two additional subsets S_4 and S_5 of U, it is clear that no system of distinct representatives exists. Further, there are exactly three systems of distinct representatives of S_1, S_2, and S_3, namely $(1, 2, 3)$, $(1, 2, 4)$, and $(1, 3, 4)$.

In general, if $S_1, S_2, \ldots, S_n$ are subsets of U, then $(s_1, s_2, \ldots, s_n)$ is a **system of distinct representatives** of the sets $S_1, S_2, \ldots, S_n$ if $s_i \in S_i$ and $s_i \neq s_j$, $i \neq j$.

The following important theorem was first proved by Philip Hall in 1935.

THEOREM 1   Let $S_1, S_2, \ldots, S_n$ denote subsets of U. A necessary and sufficient condition for the existence of a system of distinct representatives of the sets $S_1, S_2, \ldots, S_n$ is that for every $k = 1, 2, \ldots, n$, and every choice $S_{i_1}, S_{i_2}, \ldots, S_{i_k}$ of k of the given subsets, the union

$$S_{i_1} \cup S_{i_2} \cup \cdots \cup S_{i_k}$$

of the sets has cardinality at least equal to k.

The proof can be given by induction. For a proof, see, for example, "Combinatorial Theory" by Marshall Hall, Jr., (Blaisdell, Waltham, Massachusetts, 1967).

If there is at least one system of distinct representatives, there may be many. The following theorem, also given without proof, is a corollary of Theorem 1.

THEOREM 2   If the sets $S_1, S_2, \ldots, S_n$ have a system of distinct representatives, and if the smallest cardinality of these sets is k, then if $k \leq n$, there are at least $k!$ different systems of distinct representatives and if $k > n$ there are at least $k^{(n)}$.

EXAMPLE 2   The sets S_2 and S_3 of Example 1 have the system of distinct representatives (1, 3). Thus by Theorem 2 they have at least two such systems. It is easy to see that there are exactly five different systems of distinct representatives of S_2, S_3.

An example of the application of systems of distinct representatives to latin squares (defined in Section 16.2) is suggested in Problem (5) of Section 16.2.

Systems of distinct representatives are closely connected to matchings in bipartite graphs. This is illustrated in the following example.

EXAMPLE 3   Let S_1, S_2, S_3 be the sets of Example 1. Let the elements of $U = \{1, 2, 3, 4\}$ be represented by the vertices of a graph and let the sets S_1, S_2, S_3 be also represented by vertices, with the vertex representing element i joined to the vertex representing the set S_j if i is an element of S_j (Figure 16.1). Notice that the vertices of this graph can be partitioned into

Figure 16.1

two sets such that each edge has a vertex in each set. Such a graph is called a **bipartite graph**. A subset of the edges such that no vertex is on two of the edges is called a **matching** of the bipartite graph. A system of distinct representatives is represented by a set of edges which is a matching. Such a matching is a **maximal matching** (no edges can be added) having the property that there is an edge associated with each vertex S_1, S_2, S_3. But it should be noted that not every maximal matching corresponds to a system of distinct representatives. For example, the set of edges (1, S_2) and (3, S_3) in Figure 16.1 is a maximal matching which does not correspond to a distinct set of representatives in Example 3.

EXERCISE 16.1

1.   Find six systems of distinct representatives of the sets $\{1, 2, 3\}$, $\{2, 3, 4\}$, $\{3, 4, 5\}$, $\{4, 5, 1\}$, $\{5, 1, 2\}$.

2.   Represent the sets in Problem 1 and the elements of $U = \{1, 2, 3, 4, 5\}$ by vertices of a bipartite graph. Find six matchings of this bipartite graph.

3.   Sets S_1, S_2, S_3, S_4, and S_5 have cardinality 2, 3, 4, 5, and 6, respectively. Show that there are at least 32 systems of distinct representatives of these sets. Generalize to n sets.

4.^{c} Write a computer program for finding systems of distinct representatives.

5.^{c} Check your program of Problem 4^c by finding systems of distinct representatives in (a) Problem 1, (b) Problem 3, with $S_i = \{1, 2, \ldots, i\}$ for $i = 2, 3, \ldots, 6$.

16.2 Latin Squares

The array

$$L = [a_{ij}] = \begin{bmatrix} a_{11} & a_{12} & \cdots & a_{1n} \\ a_{21} & a_{22} & \cdots & a_{2n} \\ \vdots & \vdots & & \vdots \\ a_{n1} & a_{n2} & \cdots & a_{nn} \end{bmatrix}$$

is called a **latin square** of order n if each row and column is a permutation of the numbers $1, 2, \ldots, n$. A **latin rectangle** is a rectangular array of r rows and s columns consisting of integers from 1 to n ($n \geq r, n \geq s$) with no integers repeated in a row or column.

Theorem 1 asserts the existence of latin squares of all orders. In the proof of this existence we use the idea of congruence from number theory.

THEOREM 1   A latin square exists for every positive integer n.

Proof Consider $[a_{ij}]$ with $a_{ij} \equiv i + j \pmod{n}$, $1 \leq a_{ij} \leq n$. Then $a_{ij} = a_{ik}$ implies $i + j \equiv i + k \pmod{n}$ and so $j \equiv k \pmod{n}$ and this means $j = k$ since i, j, and k represent positive integers $\leq n$. Similarly, $a_{ij} = a_{kj}$ implies $i = k$. Thus the elements of each row and each column are distinct and the array $[a_{ij}]$ is a latin square.

EXAMPLE 1   The latin square of order four produced by the method of Theorem 1 is

$$\begin{bmatrix} 2 & 3 & 4 & 1 \\ 3 & 4 & 1 & 2 \\ 4 & 1 & 2 & 3 \\ 1 & 2 & 3 & 4 \end{bmatrix}.$$

A latin square of order n is said to be in **normalized** form if the first row and the first column are both the ordered set $\{1, 2, \ldots, n\}$. Interchanging columns in a latin square yields a latin square. In Example 1 we obtain the normalized latin square

$$\begin{bmatrix} 1 & 2 & 3 & 4 \\ 2 & 3 & 4 & 1 \\ 3 & 4 & 1 & 2 \\ 4 & 1 & 2 & 3 \end{bmatrix}$$

by interchanging the fourth column successively with the third, second, and first columns. A second normalized latin square of order four is

$$\begin{bmatrix} 1 & 2 & 3 & 4 \\ 2 & 1 & 4 & 3 \\ 3 & 4 & 2 & 1 \\ 4 & 3 & 1 & 2 \end{bmatrix}.$$

It has been shown that there are 535,281,401,856 normalized latin squares of order eight. The reader is advised against verifying this result by the method of exhaustion without the aid of a large computer.

Two latin squares, $[a_{ij}]$ and $[b_{ij}]$ of order n, are **orthogonal** if the n^2 ordered pairs (a_{ij}, b_{ij}) are all different. Note that two normalized latin squares are not orthogonal.

Orthogonal latin squares were first studied by Euler who was unable to solve the following problem. "Given 36 officers from six ranks and six regiments, can they be arranged in a square so that each row and each column contains exactly one officer of each rank and exactly one officer of each regiment?" This is seen to be equivalent to asking for a pair of orthogonal latin squares of order six.

Euler conjectured that there did not exist pairs of orthogonal squares if n is twice an odd integer. In 1900 Tarry proved this conjecture for $n = 6$ by listing all latin squares of order six. In 1960 Bose and Shrikhande disproved Euler's conjecture by constructing a pair of orthogonal latin squares of order 22. More recently orthogonal latin squares of order 10 have been constructed.

For some values of n, more than two latin squares exist which are orthogonal in pairs. In 1923 MacNeish proved that if $n = p^k$ for some prime p, then there exist $n - 1$ latin squares orthogonal in pairs. A set of $n - 1$ mutually orthogonal latin squares of order n is said to form a **complete set**. It has been shown that, if $n \geq 3$, such a complete set is equivalent to an affine plane of order n.

Nobody knows whether or not there exist three mutually orthogonal latin squares of order ten. E. T. Parker has found approximately one million latin squares orthogonal to a particular latin square of order ten, but no two of these are orthogonal.

We now prove two simple theorems concerning latin squares, an existence theorem, and a nonexistence theorem. We use the result that a finite field of order p exists.

THEOREM 2   If p is a prime, there exist $p - 1$ latin squares orthogonal in pairs.

Proof   Consider the $p - 1$ latin squares

$$L^k = [a_{ij}^k], \qquad k = 1, 2, \ldots, p - 1,$$

where

$$a_{ij}^k \equiv i + kj \pmod{p}.$$

For each k it is easy to verify that L^k is a latin square. Consider two latin squares L^λ and L^μ, $\lambda \neq \mu$, generated in this way.

Let $x = a_{ij}^\lambda$ and $y = a_{ij}^\mu$, that is,

$$x = i + \lambda j, \qquad y = i + \mu j.$$

Then $(\lambda - \mu)j = x - y$ and $j = (\lambda - \mu)^{-1}(x - y)$; an element $(\lambda - \mu)^{-1}$ will exist such that $(\lambda - \mu)(\lambda - \mu)^{-1} \equiv 1 \pmod{p}$ since the elements are in the field $GF(p)$. Also $i = x - \lambda j = x - \lambda(\lambda - \mu)^{-1}(x - y)$. Then i and j are uniquely determined by x, y, μ, and λ and every pair (x, y) occurs for only one position (i, j) in the pair of arrays L^λ, L^μ. Hence these latin squares are orthogonal for every pair (λ, μ), $\lambda \neq \mu$.

We now use Dirichlet's drawer principle to prove a nonexistence theorem.

THEOREM 3   For any positive integer n there do not exist n latin squares orthogonal in pairs.

Proof   First note that if a permutation is performed on the numbers in a latin square L, a new latin square L' is obtained. Further if L is orthogonal to the latin square M, then L' is also orthogonal to M. Thus if there are n latin squares orthogonal in pairs, they can be transformed so that the first row of each is the permutation $(1, 2, \ldots, n)$. Consider the first element in the second row of each latin square. This element cannot be 1. Since there are $n - 1$ choices for this element, two of the squares must have the same integer k in the first position of the second row. But they already have the common pair (k, k) in the kth position of the first row. This contradiction proves the assertion.

EXERCISE 16.2

1. Construct four latin squares of order five which are orthogonal in pairs. Put them into the form with 1, 2, 3, 4, 5 occurring in the first row.

2. Let $L^k = [a_{ji}^k]$ where $a_{ij}^k \equiv i + jk$ (mod 9). Which of the arrays L^k, $k = 1$, 2, ..., 8 are latin squares? Are L^2 and L^5 orthogonal?

3. Consult the literature for examples of orthogonal latin squares of order ten.

4. Let $[a_{ij}]$ denote a latin square, let

$$R_i^k = \{a_{ij}, \quad j = 1, 2, \ldots, k\}$$

and let S_i^k denote the complement of R_i^k relative to $U = \{1, 2, \ldots, n\}$, $k = 1, 2, \ldots, n - 1$. Prove that the lth column of $[a_{ij}]$ is a system of distinct representatives of the sets $S_i^k, i = 1, 2, \ldots, n$, for every $l = k + 1, \ldots, n$.

5. Use the result in Problem 4 to construct a latin square of order four with the aid of a system of distinct representatives.

6.^{c} Write a computer program to construct latin squares of any order.

7.^{*c} Let l_n be the number of normalized latin squares of order n. Find l_n for $n \le 5$.

8.* Extend Theorem 1 to the existence of $p^n - 1$ latin squares by replacing the field of integers modulo p ($GF(p)$) by $GF(p^n)$.

9.* Let $n = p_1^{\alpha_1} \cdots p_n^{\alpha_n}$ and $t = \min_i(p_i^{\alpha_i} - 1)$. Show that for $t \ge 2$ there exists a set of t orthogonal latin squares. In particular, if n is odd, there exist at least two orthogonal latin squares of order n. *Hint*: Consult H. Ryser, "Combinatorial Mathematics," p. 84, Mathematical Association of America, Providence, Rhode Island, 1965.

16.3 The Kirkman Schoolgirl Problem

In 1850, T. P. Kirkman suggested the following problem. "A schoolmistress took her fifteen girls for a daily walk, the girls arranged in five rows of three girls so that each girl had two companions. Plan the walk for seven consecutive days so that no girl walks with any of her classmates in any triplet more than once."

There are 845 solutions to this problem. One of the solutions is shown in Table 16.1.

TABLE 16.1

Sunday			Monday			Tuesday			Wednesday			Thursday			Friday			Saturday		
1	2	3	1	4	5	1	6	7	1	8	9	1	10	11	1	12	13	1	14	15
4	8	12	2	8	10	2	9	11	2	12	14	2	13	15	2	4	6	2	5	7
5	10	15	3	13	14	3	12	15	3	5	6	3	4	7	3	9	10	3	8	11
6	11	13	6	9	15	4	10	14	4	11	15	5	9	12	5	11	14	4	9	13
7	9	14	7	11	12	5	8	13	7	10	13	6	8	14	7	8	15	6	10	12

In the general problem we wish to arrange n girls (n an odd multiple of 3) in triplets to walk for $d = \frac{1}{2}(n - 1)$ days so that no girl walks with any of her classmates in any triplet more than once.

Many methods have been developed for special cases of this problem (see, for example, "Mathematical Recreations" by W. W. Rouse Ball, revised by H. S. M. Coxeter, MacMillan, London, 1967). In this section we will illustrate one of these methods for $n = 9$. This method can be generalized to any integer n of the form $24m + 3$ or $24m + 9$.

Denote the nine girls by the integers 1, 2, ..., 9. Place the 9 at the center of a circle and the integers 1, 2, ..., 8 equally spaced around the circumference. Consider the diameter 1 9 5 of the circle and the two triangles 2 3 8 and 6 7 4 as shown in Figure 16.2.

These three triplets give a suitable arrangement of the nine girls into three triplets for one day. Now consider the possible rotations of the configuration through angles of 45°, 90°, and 135° counterclockwise about the center but leaving the numbers fixed. We obtain four distinct positions (counting the

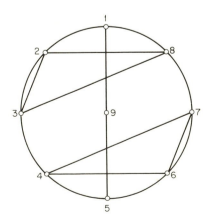

Figure 16.2

original one) for the diameter and the associated triangles. Further all the triangles obtained are distinct. This geometrical property is what we require to solve the given problem. The four positions of the configuration yield suitable arrangements.

It is clear that the construction requires the existence of scalene triangles (the triples) which cannot be rotated into each other. These can be shown to exist for certain values of *n*.

E X E R C I S E 16.3

1. Verify that Table 16.1 gives a solution to the Kirkman schoolgirl problem for *n* = 15.

2. Write down the solution to the nine-girl problem which is implied by Figure 16.1.

3. Construct a different pair of triangles which will yield another solution for the nine-girl problem.

4.* Verify that for 33 girls a suitable arrangement for the first day is (33, 1, 17), (2, 11, 16),   (4, 6, 10),   (5, 13, 30),   (7, 8, 19),   (9, 28, 31),   (18, 27, 32), (20, 22, 26), (21, 19, 14), (23, 24, 3), (25, 52, 15) by drawing a circle with center labeled 33 and taking integers 1, 2, ..., 32 equally spaced around the circumference so that the triples are represented by a diameter and ten scalene triangles.

16.4 Balanced Incomplete Block Designs

Designs were first studied by statisticians in connection with the subject of the design of experiments which, through the work of R. A. Fisher and his followers, has become an important part of modern statistical theory. The terminology that is normally used in this area has its origins in the applications of the theory rather than in the abstract formulation which can easily be made.

We begin with a set of v objects $v_1, v_2 \ldots, v_v$, called **varieties** (in the early applications these were plants or schoolgirls or fertilizers, etc.). We will be concerned with the possible arrangements of the v elements rather than with their physical properties in the applications.

Each variety is taken r times to form a collection of vr objects (or marks). These are distributed into b subsets $B_1, B_2, \ldots, B_b$ called *blocks*, each containing k varieties. No block contains a variety more than once. Clearly

$$vr = bk \qquad (1)$$

Not every such system of blocks is a required design. We also insist that each pair of varieties (v_i, v_j) occurs in exactly λ blocks. There are exactly $v(v-1)/2$ pairs of varieties and each block contains $k(k-1)/2$ pairs of varieties. Since every variety occurs in exactly r blocks, it occurs in $r(k-1)$ pairs. Summing over all varieties, we must have

$$v\frac{r(k-1)}{2} = \lambda\frac{v(v-1)}{2},$$

since each pair contains two varieties and occurs λ times, and so

$$r(k-1) = \lambda(v-1). \tag{2}$$

Another way of checking this result is by setting up a pairwise incidence matrix

$$
\begin{array}{c}
\\
v_1v_2 \\
v_1v_3 \\
\vdots \\
\\
v_{v-1}v_v
\end{array}
\begin{array}{c}
B_1 \quad B_2 \quad \cdots \quad B_b \\
\left[
\begin{array}{cccc}
a_{11} & a_{12} & \cdots & a_{1b} \\
a_{21} & a_{22} & \cdots & a_{2b} \\
\vdots & \vdots & & \vdots \\
a_{i1} & a_{i2} & \cdots & a_{ib} \\
\vdots & \vdots & & \vdots
\end{array}
\right]
\end{array}
$$

in which the ijth entry is 1 if the ith pair of varieties occurs in block j and 0 otherwise. Now count the total number of 1's first along the rows and then along the columns. Each pair occurs in exactly λ blocks so that each row contains λ 1's. Also the number of pairs of varieties is $v(v-1)/2$ so that the number of 1's in the matrix is $\lambda v(v-1)/2$. Now each block contains $k(k-1)/2$ pairs and so each column of the matrix contains $k(k-1)/2$ 1's. Since there are b columns, the matrix contains $bk(k-1)/2$ 1's. Hence

$$\frac{\lambda v(v-1)}{2} = \frac{bk(k-1)}{2} = \frac{vr(k-1)}{2}$$

and so

$$\lambda(v-1) = r(k-1) \quad \text{as before.}$$

A system of blocks satisfying (1) and (2) is called a balanced incomplete block design. A **balanced incomplete block design** or **(v, k, λ)-block design** is an arrangement of v varieties into b blocks such that each block contains exactly k varieties and every pair of varieties occurs together in exactly λ blocks.

EXAMPLE 1   Let the varieties be the integers 0, 1, ..., 6. Then

$$0 \ 1 \ 3$$
$$1 \ 2 \ 4$$
$$2 \ 3 \ 5$$
$$3 \ 4 \ 6$$
$$4 \ 5 \ 0$$
$$5 \ 6 \ 1$$
$$6 \ 0 \ 2$$

is a balanced incomplete block design with $v = 7$, $b = 7$, $k = 3$, $r = 3$, and $\lambda = 1$. Notice that this design is equivalent to the Fano plane (Section 11.2) if we interpret varieties to be points and blocks to be lines.

EXAMPLE 2   The following design is obtained by identifying opposite pairs of points on an icosahedron, labeling them 1, 2, ..., 6, and enumerating the ten resulting triangles. (Any relabeling of the six points gives an equivalent design.)

$$1 \ 2 \ 6 \qquad 1 \ 3 \ 4 \qquad 3 \ 6 \ 4 \qquad 2 \ 4 \ 5 \qquad 5 \ 1 \ 6$$
$$2 \ 3 \ 6 \qquad 1 \ 2 \ 4 \qquad 6 \ 4 \ 5 \qquad 2 \ 3 \ 5 \qquad 1 \ 3 \ 5$$

In this case, $v = 6$, $b = 10$, $k = 3$, $r = 5$, and $\lambda = 2$.

In Example 1, λ has the value 1 and the design is a projective plane. However $\lambda = 1$ does not imply that the design *is* a projective plane, unless $v = b$.

If $\lambda = 1$ and $k = 3$, that is, if each block has exactly three varieties, then the design is called a **Steiner triple system**. Thus the Fano plane is an example of a Steiner triple system.

It can be shown that a necessary and sufficient condition for the existence of a Steiner triple system is $v \equiv 1$ or $3 \pmod 6$. Steiner posed the problem of the existence of these triple systems in 1853 not knowing about Kirkman's solution to the problem in 1847. Any solution to the Kirkman schoolgirl problem is a Steiner triple system.

EXERCISE 16.4

1. Show that the following system of blocks is a design and evaluate the parameters b, v, k, r, and λ.

$$1 \ 2 \ 3 \qquad 1 \ 4 \ 7 \qquad 1 \ 5 \ 9 \qquad 1 \ 6 \ 8$$
$$4 \ 5 \ 6 \qquad 2 \ 5 \ 8 \qquad 2 \ 6 \ 7 \qquad 2 \ 4 \ 9$$
$$7 \ 8 \ 9 \qquad 3 \ 6 \ 9 \qquad 3 \ 4 \ 8 \qquad 3 \ 5 \ 7$$

2. The incidence matrix $A = (a_{ij})$ of a design is defined by

$$a_{ij} = \begin{cases} 1 & \text{if} \quad v_i \in B_j \\ 0 & \text{otherwise.} \end{cases}$$

Show that A and its transpose A^{T} obey

$$A \cdot A^{\mathrm{T}} = \begin{bmatrix} r & \lambda & \cdots & \lambda \\ \lambda & r & \cdots & \lambda \\ \vdots & \vdots & & \vdots \\ \lambda & \lambda & \cdots & r \end{bmatrix}$$

$$= (r - \lambda)I_v + \lambda J_v$$

where I_v is a unit matrix of order v and J_v is a matrix of order v all of whose entries are 1's.

3.   Show that $\det(AA^{\mathrm{T}}) = (r - \lambda)^{v-1}[r + (v - 1)\lambda]$.

4.   Show that $JA = kJ$, where J is defined as in Problem 2.

5.   Verify that Table 16.1 of Section 16.3 gives a Steiner triple system.

6.   Verify that the following twelve blocks on nine varieties is a Steiner triple system.

1 2 3	1 7 9	2 7 8	3 6 9
1 4 5	2 4 9	3 4 8	4 6 7
1 6 8	2 5 6	3 5 7	5 8 9

7.*   In a school made up of b boys, t athletic teams are formed. The athletic director puts k boys on each team, and arranges that every boy plays on the same number of teams. He also arranges that each pair of boys plays together the same number of times. Find how many teams a boy plays on and how often two boys play on the same team. (Ontario Senior Mathematics Problems Competition, 1968.)

16.5 Difference Sets

Difference sets are illustrated in the following example.

EXAMPLE 1   Instead of denoting the points of the Fano plane (Section 11.2) by $F_1, F_2, \ldots, F_7$, we use the integers modulo 7, as in Section 16.4. The points of the geometry are then represented by the integers 0, 1, 2, ..., 6, and the lines are obtained from the triple (0, 1, 3) by successively adding the integers modulo 7. This gives us the required set of lines because the differences between the integers 0, 1, and 3 are ± 1, ± 2, and ± 3 and these are all the nonzero residues modulo 7.

A set of residues $(d_1, d_2, \ldots, d_k)$ is called a **difference set modulo** v, or a (v, k, λ)-**difference set** if the differences $d_i - d_j$, $i \neq j$, contain all of the nonzero residues modulo v, and each difference occurs the same number, λ, of times. It is clear that

$$k(k - 1) = \lambda(v - 1). \tag{1}$$

The following theorems are given without proof.

THEOREM 1   Let $(d_1, d_2, \ldots, d_k)$ be a $(v, k, 1)$-difference set. Let $\mathscr{P}$ denote the set of residues modulo v and let $\mathscr{L}$ denote the collection of sets L_j for $j = 0, 1, 2, \ldots, v - 1$ where

$$L_j = \{j + d_i \,(\text{mod } v); \quad i = 1, 2, \ldots, k\}.$$

Then $(\mathscr{P}, \mathscr{L}, I)$ is the projective plane $PG(k - 1)$ where incidence I denotes set inclusion.

THEOREM 2   If $(d_1, d_2, \ldots, d_k)$ is a (v, k, λ)-difference set, then $B_0, B_1, \ldots,$ B_{v-1} is a (v, k, λ)-block design where

$$B_j = \{j + d_i \,(\text{mod } v); \quad i = 1, 2, \ldots, k\}$$

for $j = 0, 1, \ldots, v - 1$. (The elements of B_j are residue classes modulo v.)

EXAMPLE 2   The residues $(0, 1, 2, 4, 5, 8, 10)$ modulo 15 form a $(15, 7, 3)$-difference set since each of the 14 nonzero residues modulo 15 occurs three times among the 42 differences

$$\pm\{1, 2, 4, 5, 8, 10, 1, 2, 4, 7, 9, 2, 3, 6, 8, 1, 4, 6, 3, 5, 3\}$$

(the $\pm$ is to be associated with each element of the set). Hence the blocks

$$B_j = \{j, j + 1, j + 2, j + 4, j + 5, j + 8, j + 10\},$$

$j = 0, 1, \ldots, 14$, form a $(15, 7, 3)$-block design. The fifteen blocks are as follows.

B_0	0	1	2	4	5	8	10
B_1	1	2	3	5	6	9	11
B_2	2	3	4	6	7	10	12
B_3	3	4	5	7	8	11	13
B_4	4	5	6	8	9	12	14
B_5	5	6	7	9	10	13	0
B_6	6	7	8	10	11	14	1
B_7	7	8	9	11	12	0	2
B_8	8	9	10	12	13	1	3
B_9	9	10	11	13	14	2	4
B_{10}	10	11	12	14	0	3	5
B_{11}	11	12	13	0	1	4	6
B_{12}	12	13	14	1	2	5	7
B_{13}	13	14	0	2	3	6	8
B_{14}	14	0	1	3	4	7	9

This block design can be interpreted as the fifteen planes of the three dimensional projective geometry $PG(3, 2)$, with the fifteen residues representing the points. Each plane is a Fano plane $PG(2, 2) \equiv PG(2) \equiv PG(GF(2))$.

All the lines of the geometry can be obtained as follows. The plane B_j contains the line

$$L_j = \{j, j + 1, j + 4\}.$$

All the lines of B_j may be obtained by cyclically permuting the points of L_j according to the permutation

$$B_j' = (j, j + 1, j + 2, j + 4, j + 5, j + 10, j + 8).$$

EXAMPLE 3   The lines in the plane

$$B_0 = \{0, 1, 2, 4, 5, 10, 8\}$$

are

$$(0, 1, 4), (1, 2, 5), (2, 4, 10), (4, 5, 8),$$
$$(5, 10, 0), (10, 8, 1), (8, 0, 2).$$

The lines in the plane

$$B_3 = \{3, 4, 5, 7, 8, 13, 11\}$$

are

$$(3, 4, 7), (4, 5, 8), (5, 7, 13), (7, 8, 11),$$
$$(8, 13, 3), (13, 11, 4), (11, 3, 5).$$

EXERCISE 16.5

1. Show that $(0, 1, 3, 9)$ is a simple difference set modulo 13 (each residue occurs as a difference exactly once).

2. Show that if the set $\mathscr{P}$ of residues modulo 13 is taken as a set of points, and if the sets

    $$L_j = \{0 + j, 1 + j, 3 + j, 9 + j\}, \qquad j = 0, 1, \ldots, 12,$$

    are taken as the set $\mathscr{L}$ of lines, and if I denotes inclusion of points in lines, then $(\mathscr{P}, \mathscr{L}, I)$ is the finite geometry $PG(3)$.

3. Show that $(0, 1, 5)$ is a difference set modulo seven which yields $PG(3)$ as in Example 1.

4. If $(d_1, d_2, \ldots, d_k)$ is a difference set modulo v, then

    $$(d_1 + j, d_2 + j, \ldots, d_k + j)$$

    is also a difference set for every j.

5. List the 35 lines of the projective geometry $PG(3, 2)$ described in Example 2.

6.* As in Example 2, verify that

$$D = (0, 1, 2, 32, 33, 12, 24, 29, 5, 26, 27, 22, 18)$$

is an (ordered) difference set modulo 40. Determine the parameters v, k, λ for this difference set.

7.* Show that if $(0, 1, 26, 32)$ is taken as the line L_0 then twelve additional lines are obtained by performing successively on L_0 the cyclic permutation determined by D in Problem 6, (for example, $L_1 = (1, 2, 27, 33)$, $L_2 = (2, 32, 22, 12)$, etc.). The thirteen points $0, 1, 2, 32, 33, \ldots, 22, 18$ of Problem 6 together with the thirteen lines L_i, $i = 0, 1, \ldots, 12$, form a projective plane $PG(3)$. (See Problem 2.)

16.6 Magic Squares

A **magic square** is a square array of integers with the property that the sum of the integers in every row, column, and diagonal is the same. In particular, if the integers are $1, 2, \ldots, n^2$, the square is said to have order n.

EXAMPLE 1   Two 3×3 magic squares are shown in Figure 16.3. The sums are 111 in Figure 16.3a and 15 in Figure 16.3b. The latter square has order three.

67	1	43
13	37	61
31	73	7

(a)

8	1	6
3	5	7
4	9	2

(b)

Figure 16.3

THEOREM 1   If a magic square is of order n, then its (row, column, diagonal) sums are $\frac{1}{2}n(n^2 + 1)$.

Proof   Let the sum be S. Then adding *all* the rows we have

$$nS = 1 + 2 + \cdots + n^2 = \frac{1}{2}n^2(n^2 + 1).$$

Thus

$$S = \frac{1}{2}n(n^2 + 1).$$

This theorem seems to be about the only simple theorem that can be proved about magic squares. The theory consists of numerous special methods for constructing small classes of magic squares. Magic squares were studied in antiquity in China and were introduced into Europe in the fifteenth century. An interesting comprehensive account of magic squares is given in "Mathematical Recreations and Essays," W. W. Rouse Ball, revised by H. S. M. Coxeter, MacMillan, London, 1967.

To illustrate the type of constructions in the literature we include the method for constructing magic squares of odd order given by S. de la Loubère (Du Royaume de Siam, English translation, London 1693, Vol. II pp. 227–247). According to Ball, de la Loubère was the envoy of Louis XIV to Siam in 1687–1688 and there learned the following method.

De la Loubère's Method for Constructing Magic Squares

Place a 1 in the middle square of the top row. The successive integers are then placed in their natural order in a diagonal line which slopes upwards to the right with the following modifications.

(i)   When the top row is reached, the next integer is placed in the square in the bottom row as if it came immediately above the top row.

(ii)   When the right-hand column is reached, the next integer is placed in the left-hand column as if it immediately succeeded the right-hand column.

(iii)   When a square already contains a number or when the top right-hand square is reached, the next integer is placed vertically below it in the next row.

EXAMPLE   We illustrate de la Loubère's construction for $n = 5$ in Figure 16.4. The reader may check the construction for $n = 3$ in Figure 16.3b.

17	24	1	8	15
23	5	7	14	16
4	6	13	20	22
10	12	19	21	3
11	18	25	2	9

Figure 16.4

EXERCISE 16.6

1.   Verify that the array in Figure 16.4 is a magic square.

2.   Construct a magic square of side seven using de la Loubère's method.

3.   Consult Ball's book for other methods of constructing magic squares, in particular, magic squares of even order.

16.7 Room Squares

Room squares were introduced by T. G. Room in an article entitled "A New Type of Magic Square" in the *Mathematics Gazette*, 1955. However Room squares were first introduced into duplicate bridge under the name of Howell Movements by E. C. Howell about 1897.

A **Room square** of side n is an $n \times n$ array of cells, n odd, in which each cell is either empty or contains an unordered doublet of symbols from the set $\{\infty, 0, 1, 2, \ldots, n - 1\}$ with the following properties holding:

(i)   The array is "latin" by columns (every symbol occurs exactly once in a column).
(ii)   The array is latin by rows.
(iii)   All lines contain exactly $\frac{1}{2}(n - 1)$ blank cells.
(iv)   All doublets are distinct (each possible doublet occurs once and only once).

EXAMPLE   In Figure 16.5 we exhibit a Room square of side seven. The reader should check that this array has all the properties stipulated for a Room square.

∞ 0	62	45		13		
	∞ 1	03	56		24	
		∞ 2	14	60		35
46			∞ 3	25	01	
	50			∞ 4	36	12
23		61			∞ 5	40
51	34		02			∞ 6

Figure 16.5

If $n = 2k + 1$, the number of possible doublets is

$$\binom{2k + 2}{2} = \tfrac{1}{2}(2k + 1)(2k + 2)$$

and the number of filled cells is $(k + 1)(2k + 1)$ since each line contains k blank cells.

The question arises as to whether or not Room squares exist for all odd positive integers. Certainly $\infty 0$ is a Room square of side one and we have seen in the above example a Room square of side seven. The question is partially answered by the result in the following theorem.

THEOREM   There does not exist a Room square of side three.

Proof   In the following we will use a_{ij} to indicate both the cell in row i and column j and the doublet in that cell if the cell is not empty.

The symbols ∞ and 0 must occur somewhere as a doublet. Interchange rows and columns if necessary so that $a_{11} = \infty 0$ (Figure 16.6). Now the doublet 12 must occur in row 1 and also in column 1. But $a_{11} \neq 12$ and the doublet cannot occur twice. We conclude that a Room square of side three cannot exist.

Figure 16.6

At the present time, three and five are the only odd integers for which it is known that Room squares do not exist. The reader is asked to verify in Exercise 16.7 that a Room square of side five does not exist. The method of exhaustion is used.

It has been conjectured that Room squares exist for all sides other than 3 or 5. This has now been verified (1971) for all odd integers other than 257 and its multiples. A multiplication theorem, that Room squares of sides m and n can be compounded to give a square of side mn, has been proved.[1] The computer has been used extensively in producing Room squares.

EXERCISE 16.7

1.   Use the method of exhaustion to prove that a Room square of side five does not exist. Proceed as follows.

   (i)   The first row has three cells filled and two empty. Interchange columns and permute symbols if necessary so that $a_{11} = \infty 0$, $a_{12} = 12$, and $a_{13} = 34$.

[1] R. G. Stanton and R. C. Mullin, Construction of Room squares, *Ann, Math. Stat.* **39** (1968), 1540–1548.

(ii) The remaining doublets in the first column must be either 13 and 24 or 14 and 23. Permute 3 and 4 in this column and interchange rows if necessary so that $a_{21} = 13$ and $a_{31} = 24$. Then a_{14}, a_{15}, a_{41}, and a_{51} are empty cells (Figure 16.7).

Figure 16.7

(iii) Show that the doublets 14 and 23 must be in rows 4 or 5 and columns 4 or 5 and show that we may take $a_{44} = 14$ and $a_{55} = 23$.

(iv) Show that a_{45} and a_{54} must be empty.

(v) Now rows 4 and 5 and columns 4 and 5 already have two blank cells so that the remaining cells in these rows and columns must be filled. Then a_{22}, a_{23}, a_{32}, and a_{33} must be empty cells.

(vi) Cell a_{42} is either $\infty 3$ or 03. Rename symbols if necessary so that $a_{42} = 03$. Then $a_{52} = \infty 4$, $a_{43} = \infty 2$, and $a_{53} = 01$.

(vii) Now a_{35} must involve ∞, 0, or 1, that is, must be one of the doublets $\infty 0$, $\infty 1$, or 01. But $\infty 0$ and 01 already appear, so that $a_{35} = \infty 1$ and then $a_{25} = 04$.

(viii) Show now that it is impossible to allocate doublets to the cells a_{24} and a_{34} and hence conclude that a Room square of side five does not exist.

2. Repeat Problem 1 by making the same steps (i) and (ii) but at step (iii) choosing $a_{44} = 25$ and $a_{55} = 34$. Complete the argument.

3.*ᶜ Write a computer program to find Room squares and check your program for a square of side seven.

ANSWERS TO SELECTED EXERCISES

Chapter 1

4. $2^k - 1$ **5.** 585   **6.** 3, 689, 348, 814, 742

2. (a) 6   (b) 10   (c) 15   **3.** $\dfrac{n(n+1)}{2}$

4. 6   **5.** 10   **6.** 13

7. $(n+1)k + \dfrac{n(n+1)}{2} + 1$

2. (b) 60; 60; 5

EXERCISE 1.4

2. M_1: 2; 6; 12
M_2: 0; 6; 24
M_3: 2; 18; 24
M_4: 2; 12; 24

3. $\lambda^5 - 5\lambda^4 + 10\lambda^3 - 10\lambda^2 + 4\lambda$

EXERCISE 1.6

1. (a) (i) 260; 630; 1302; 2408
(ii) 1949; 4032; 7449; 12,674
(iii) 112; 200; 324; 490

4. (c) If $f(x)$ is a monic polynomial of degree n, then $\Delta^n f(x) = n!$

EXERCISE 1.7

1. For $k = 3$: d^3; $3pd^2$; $3p^2d$; p^3
For $k = 4$: d^4; $4pd^3$; $6p^2d^2$; $4p^3d$; p^4

2. $d^5 + 5pd^4 + 10p^2d^3$

3. (a) $\dfrac{560}{2187}$ (b) $\dfrac{28}{729}$

4. $(2, 0)$

Chapter 2

EXERCISE 2.1

1.

θ	I	P_1	P_2	P_3	P_4	P_5
I	I	P_1	P_2	P_3	P_4	P_5
P_1	P_1	I	P_3	P_2	P_5	P_4
P_2	P_2	P_4	I	P_5	P_1	P_3
P_3	P_3	P_5	P_1	P_4	I	P_2
P_4	P_4	P_2	P_5	I	P_3	P_1
P_5	P_5	P_3	P_4	P_1	P_2	I

2. (a) $\begin{pmatrix} 1 & 2 & 3 \\ 1 & 3 & 2 \end{pmatrix} \begin{pmatrix} 1 & 2 & 3 \\ 2 & 1 & 3 \end{pmatrix}$

(b) $\begin{pmatrix} 1 & 2 & 3 & 4 \\ 1 & 3 & 2 & 4 \end{pmatrix}, \begin{pmatrix} 1 & 2 & 3 & 4 \\ 3 & 2 & 1 & 4 \end{pmatrix}$

(c) $\begin{pmatrix} 1 & 2 & 3 & 4 & 5 \\ 2 & 5 & 1 & 3 & 4 \end{pmatrix}, \begin{pmatrix} 1 & 2 & 3 & 4 & 5 \\ 5 & 4 & 1 & 3 & 2 \end{pmatrix}$

(d) $\begin{pmatrix} a & b & c & d & e \\ d & b & c & a & e \end{pmatrix}, \begin{pmatrix} a & b & c & d & e \\ a & e & c & d & b \end{pmatrix}$

3. (a) $\begin{pmatrix} 1 & 2 & 3 \\ 1 & 2 & 3 \end{pmatrix}, \begin{pmatrix} 1 & 2 & 3 \\ 3 & 1 & 2 \end{pmatrix}$

(b) $\begin{pmatrix} 1 & 2 & 3 & 4 \\ 2 & 1 & 4 & 3 \end{pmatrix}, \begin{pmatrix} 1 & 2 & 3 & 4 \\ 4 & 1 & 3 & 2 \end{pmatrix}$

(c) $\begin{pmatrix} 1 & 2 & 3 & 4 & 5 \\ 4 & 2 & 5 & 1 & 3 \end{pmatrix}, \begin{pmatrix} 1 & 2 & 3 & 4 & 5 \\ 1 & 2 & 3 & 4 & 5 \end{pmatrix}$

(d) $\begin{pmatrix} a & b & c & d & e \\ d & e & c & a & b \end{pmatrix}, \begin{pmatrix} a & b & c & d & e \\ a & b & c & d & e \end{pmatrix}$

4. (a) I   (b) $\begin{pmatrix} 1 & 2 & 3 & 4 \\ 4 & 3 & 1 & 2 \end{pmatrix}$

(c) $\begin{pmatrix} 1 & 2 & 3 & 4 & 5 \\ 5 & 2 & 4 & 3 & 1 \end{pmatrix}$

(d) $\begin{pmatrix} a & b & c & d & e \\ e & a & b & c & d \end{pmatrix}$

7. (a) $\begin{pmatrix} 1 & 2 & 3 \\ 3 & 2 & 1 \end{pmatrix}, \begin{pmatrix} 1 & 2 & 3 \\ 3 & 1 & 2 \end{pmatrix}$

(b) $\begin{pmatrix} 1 & 2 & 3 & 4 \\ 3 & 4 & 2 & 1 \end{pmatrix}, \begin{pmatrix} 1 & 2 & 3 & 4 \\ 4 & 1 & 3 & 2 \end{pmatrix}$

(c) $\begin{pmatrix} 1 & 2 & 3 & 4 & 5 \\ 3 & 2 & 4 & 5 & 1 \end{pmatrix}, \begin{pmatrix} 1 & 2 & 3 & 4 & 5 \\ 1 & 5 & 3 & 4 & 2 \end{pmatrix}$

(d) $\begin{pmatrix} a & b & c & d & e \\ e & a & c & b & d \end{pmatrix}, \begin{pmatrix} a & b & c & d & e \\ e & d & c & b & a \end{pmatrix}$

EXERCISE 2.2

1. 3^5; 3×4^4 **2.** (a) $9 \times 9 \times 8 \times 7 \times 6$ (b) 9×10^4

3. $8!$ **4.** 120

5. (i) 4536   (ii) 2296   (iii) 120

6. 90,360   **7.** 7   **8.** (a) 5!   (b) 72

9. 5! 4!   **10.** $\begin{pmatrix} 13 \\ 8 \end{pmatrix}$ or $\begin{pmatrix} 13 \\ 5 \end{pmatrix}$

11. (a) $4 \times 7!$ (b) $\dfrac{11!}{8}$ (c) $\dfrac{10!}{2!4!}$

12. 300   **13.** $\dfrac{17!}{2!6!9!}$ **14.** $\dfrac{17!}{2!6!9!}$

15. (a) 6   (b) 12   (c) 72   (d) 72   **16.** 47

EXERCISE 2.3

1. 24   **2.** 1632

3. (a) 2160   (b) 1440   (c) 5040

4. 3003   **5.** 980   **6.** 276; 2024

7. 770   **8.** $\begin{pmatrix} 52 \\ 13 \end{pmatrix}$

9. $\beta(\beta - 1)(\beta - 2) \cdots (\beta - \alpha + 1)$ **10.** 90

11. $\begin{pmatrix} m + n \\ n \end{pmatrix}$ **12.** $\dfrac{\binom{15}{5}\binom{10}{5}}{3!}$

13. 3^{15} **14.** 63

15. (a) 44,800   (b) $\begin{pmatrix} 20 \\ 2 \end{pmatrix}\begin{pmatrix} 16 \\ 2 \end{pmatrix}\begin{pmatrix} 14 \\ 2 \end{pmatrix}\begin{pmatrix} 10 \\ 2 \end{pmatrix}$

16. 185

EXERCISE 2.4

2. 14   **3.** 84   **4.** 78

6. $(p + q)^n$ can be expressed as the sum of all terms of the form

$$\frac{n!}{a!b!} p^a q^b,$$

where $a + b = n$, and a and b are positive integers.

8. $c_1 = na^{n-1}; c_2 = \binom{n}{2} a^{n-2}; c_i = \binom{n}{i} a^{n-i}$

11. (a) 150 (b) $\binom{7}{2} 9^5 3^2$ (c) $\binom{14}{3} 12^{11} 3^3$

13. $\binom{n}{\frac{4}{5}n}$

14. $x^6 + 6x^5 + 9x^4 - 4x^3 - 9x^2 + 6x - 1$

EXERCISE 2.5

4. $\binom{n-2}{r-2} + 2\binom{n-2}{r-1} + \binom{n-2}{r}$ **5.** 3

8. The sum of elements of nth row is 2^{n-1}

11. 1287   **12.** 210   **13.** 6

EXERCISE 2.6

4. $3^6; 4^{13}; 5^{14}$ **5.** 3^{10}

7. $126\, a^4 b^5 + 252\, a^6 bc^2 + 504\, a^5 b^3 c$

8. $\dfrac{n^2(n+1)^2}{4}$

Chapter 3

EXERCISE 3.1

2. (a) $A + B + C - AB - AC - BC + ABC$
(b) $AD - ABD - ACD + ABCD$

3. (a) $U - A$ (b) $U - A - B + AB$
(c) $U - A - B - C + AB + AC + BC - ABC$
(d) $BC - ABC - BCD + ABCD$
(e) $C - AC - BC - DC + ABC + ADC + BCD - ABCD$
(f) $AB - ABC - ABD - ABE + ABCD + ABCE + ABDE - ABCDE$
(g) $U - ADE + ACDE$

6. $A'B'C' \cdots N' = (A \cup B \cup C \cup \cdots \cup N)'$

EXERCISE 3.2

1. 5; 5   **2.** 4; 5; 4; 6

3. -58 **4.** 60

EXERCISE 3.3

1. 515   **2.** 22,400

3. 6233   **4.** 998,910

5. 12   **6.** $\lambda(\lambda - 1)(\lambda - 2)(\lambda - 3)$

7. 61

EXERCISE 3.4

1. 9   **2.** 44; 1854

3. (a) $[D(6)]^2 = 70,225$ (b) $(6!)^2 = 518,400$

4. 504   **5.** 5544

7. (a) 14,833 (b) 25,487 (c) 10,655

8. (a) 1854 (b) 3186 (c) 1

11. 14,833   **12.** $\dfrac{1}{e}$

Chapter 4

EXERCISE 4.1

1. (a) $\binom{35}{4}$ (b) $\binom{40}{4}$

4. (a) 246   (b) 194

5. 35

6. 126

7. 53,130

8. 11,628

10. 3003

11. $\binom{n+r-1}{n-1}$

13. 969

14. (a) $\binom{61}{5}$ (b) $\binom{56}{5}$

15. 105

16. $\binom{t-k_1-k_2+3}{3}$

EXERCISE 4.2

1. 231

2. 115

3. 140

4. 136

5. 29

6. 270

8. 37038

10. 35

EXERCISE 4.3

1. 220

2. 210

Chapter 5

EXERCISE 5.1

1. (a) $f(n) = f(n-1) + 2n - 2, f(1) = 2$
(b) $f(n) = f(n-1) + 2n - 2, f(1) = 2$
(c) $f(n) = (2n-1)f(n-1), f(1) = 1$
(d) $f(n) = f(n-1) + 1 + \frac{1}{2}n(n-1), f(1) = 2$
(e) $f(n, r) = f(n-1, r) + f(n-2, r-1), f(1, 1) = 1$

2. (a) $a_n = -3a_{n-1}$ (b) $a_n = a_{n-1} + a_{n-2}$ (c) $a_n = \dfrac{1}{n} a_{n-1}$

EXERCISE 5.2

1. $s_n = 2^{n+1} + 2^n - 1$

2. $f(n) = \dfrac{n(n+1)}{2} - 1$

3. $P(n, r) = n^{(r)}$

4. $t_n = \dfrac{n^2(n+1)^2}{4}$

5. $t_n = \dfrac{n(3n+1)}{2}$

6. $t_n = \dfrac{n}{4n+1}$

7. $t_n = 2^n - 1$

8. $t_n = n^2$

9. $t_n = \dfrac{n(9-n)}{2}$

10. $t_n = \dfrac{n(n+1)(n+2)(n+3)}{4}$

EXERCISE 5.3

2.   (i) (a) 1, 1, 1   (b) 0, 0, 0, 0
        (ii) (a) $-2, -1, 0$ (b) 1, 1, 1, 1
        (iii) (a) 1, 2, 5   (b) 1, 3, 5, $2x + 1$
        (iv) (a) 0, 21, 24   (b) 21, 3, $-9, 3x^2 - 21x + 21$
        (v) (a) 1, 2, 4   (b) 1, 2, 4, 2^x

3. (i) 0, 0   (ii) 0, 0   (iii) 2, 2   (iv) $-12, 6x - 18$ (v) $2, 2^x$

4. 3^{x+nh}

5. $a_n n!$ where a_n is the coefficient of x^n

9. (a) $\dfrac{2n+2}{n+2}$ (b) $n(n+2)$ (c) $n^3 + 3n^2 + 3n$

10. $\dfrac{n(n^2-1)}{3}$

EXERCISE 5.5

1. (a) $\frac{1}{2}n(n-1)+2$ (b) $\frac{n(n-1)(2n-1)}{6}+1$ (c) $\frac{n^2(n-1)^2}{4}+1$

2. (a) $\frac{n(n-1)(n-2)}{3}+1$ (b) $\frac{2n-1}{n}$ (c) n^2 (d) n^3 (e) $(n-2)\,2^n+3$

EXERCISE 5.6

4. (a) $\lambda(\lambda-1)^2(\lambda-2)$ (b) $\lambda(\lambda-1)^3$ (c) $\lambda(\lambda-1)^3$

5. (a) 252   (b) 360   (c) 96   (d) 1440   (e) 0

6. 96   **7.** 6

8. 2   **9.** 732

Chapter 6

EXERCISE 6.1

1. (a) $(1+x)^{-1}$ (b) $(1-x)^{-2}$ (c) $(1+x)^{-2}$ (d) $(1+2x)^{-1}$

  (e) $\left(1-\frac{x}{2}\right)^{-1}$ (f) $(1-5x)^{-1}$ (g) $x^2(1-x)^{-2}$ (h) e^x (i) $2x(1-x)^{-3}$

3. (a) $a_n = \frac{1}{n!}$

  (b) $a_n = 0,\ n$ even; $a_n = \frac{1}{n!},\ n = 4k+1;\ a_n = \frac{-1}{n!},\ n = 4k+3$

3. (c) $a_n = 0,\ n$ odd; $a_n = \frac{1}{n!},\ n = 4k;\ a_n = \frac{-1}{n!},\ n = 4k+2$

EXERCISE 6.2

1. $s_n = 3n + 1$ **2.** $a_n = 1 + \frac{n(n-1)}{2}$

3. $t_n = \dfrac{n(n+1)(n+2)}{6}$ **4.** $a_n = n^2$

5. $a_n = \dfrac{9n - n^2}{2}$ **6.** $a_n = 2^n - 1$

7. $y_k = (1 - B)A^k + \dfrac{B(A^{k+1} - 1)}{A - 1}$ **8.** $f(n) = \dfrac{n(n+1)(2n+1)}{6}$

9. (a) $xA(x)$ (b) $\dfrac{1}{x}[A(x) - a_0]$

EXERCISE 6.3

2. $\{a_n + b_n\}$, $\left\{\displaystyle\sum_{k=0}^{n} a_k b_{n-k}\right\}$

EXERCISE 6.4

1. $a_n = \dfrac{1}{\sqrt{5}}\left\{(4 + \sqrt{5})\left(\dfrac{1 + \sqrt{5}}{2}\right)^n - (4 - \sqrt{5})\left(\dfrac{1 - \sqrt{5}}{2}\right)^n\right\}$

2. $y_k = 2^k - k$

3. (a) $f(n, r) = f(n - 1, r - 1) + f(n - 1, r)$
  (b) $f(n, r) = \dbinom{n-1}{r-1}$
  (c) 2^{n-1}

4. $a_n = \dfrac{(2n - 2)!}{n!(n - 1)!}$

EXERCISE 6.5

2. (a) $\dfrac{2^{n+1} - 1}{n + 1}$ (b) 1, if $n = 0$; -1 if $n = 1$; 0, otherwise

3. $5e$ **4.** $f_n = (n + 1)2^n$

Chapter 7

EXERCISE 7.2

2. Yes   **6.** No

7. (a) Yes   (b) No   **8.** No

Chapter 8

EXERCISE 8.1

1. (a) Yes   (b) No   (c) Yes

EXERCISE 8.2

1. (a) Yes   (b) Yes   (c) No

2. $\dfrac{ab}{(a, b)^2}$ squares of side (a, b) where (a, b) is the greatest common divisor of a and b.

6. No   **8.** Yes

9. $\dfrac{a}{b}, \dfrac{b}{c}, \dfrac{c}{a}$, rational

EXERCISE 8.4

1. 12   **2.** 4   **3** 8.

4. (a) 8   (b) 8   **5.** 10   **6.** 20

Chapter 10

EXERCISE 10.1

1. 7; 8

2. Fig. 9.4, 1; Fig. 9.5, 2 or 3 depending on whether the number of faces is even or odd; Fig. 9.6, 4; Fig. 9.7, 3; Fig. 9.8, 4; Fig. 9.9, 2; Fig. 9.10, 3.

E X E R C I S E   10.3

1. $\lambda(\lambda - 1)(\lambda - 2)^2$ **6.** $\lambda(\lambda - 2)^n + (-1)^n \lambda(\lambda - 2)$

E X E R C I S E   10.4

6. 327

Chapter 12

E X E R C I S E   12.1

1. $\dfrac{5}{13}$

2. $\dfrac{1}{4}, \dfrac{1}{13}, \dfrac{5}{13}, \dfrac{1}{52}, 0, \dfrac{5}{52}, 0, \dfrac{31}{52}, \dfrac{5}{32}, \dfrac{7}{52}, \dfrac{9}{13}$

3. $\dfrac{1}{2}, \dfrac{2}{9}, \dfrac{1}{9}, \dfrac{7}{18}$

4. $1 - \dfrac{1}{\binom{52}{13}} \left[4\binom{39}{13} - 6\binom{26}{13} + 4 \right]$

5. $\dfrac{5}{32}$ **6.** $\dfrac{n - r + 2}{2^{r+1}}$

7. (a) $\dfrac{\binom{13}{3}\binom{39}{10}}{\binom{52}{13}}$ (b) $\dfrac{\binom{50}{11}}{\binom{52}{13}}$ (c) $\dfrac{\binom{26}{11}}{\binom{52}{13}}$

8. (a) $\dfrac{\binom{26}{4}\binom{26}{9}}{\binom{52}{13}}$ (b) $\dfrac{\binom{26}{13}}{\binom{52}{13}}$

9. $\dfrac{\binom{13}{4}\binom{13}{4}\binom{13}{3}\binom{13}{2}}{\binom{52}{13}}$

10. (a) $\dfrac{\binom{39}{7}}{\binom{52}{20}}$ (b) $\dfrac{\binom{n-b}{r-b}}{\binom{n}{r}}$

EXERCISE 12.2

1. (a) $\alpha\beta\gamma\delta + \alpha\beta\gamma D$
   (b) $\alpha\beta CD + \alpha BCD$
   (c) $\alpha\beta\gamma D + \alpha\beta CD + A\beta\gamma D + A\beta CD$
   (d) $\alpha\beta\gamma\delta + \alpha\beta\gamma D + \alpha\beta C\delta + \alpha\beta CD + \alpha B\gamma\delta + \alpha B\gamma D + \alpha BC\delta + \alpha BCD$
   (e) $\alpha\beta\gamma\delta + \alpha\beta\gamma D + \alpha B\gamma\delta + \alpha B\gamma D + A\beta\gamma\delta + A\beta\gamma D + AB\gamma\delta + AB\gamma D$

2. Inconsistent   **3.** Inconsistent

Chapter 13

EXERCISE 13.1

1. (a) $\dfrac{(n+1)(3n+2)}{2}$ (b) $\dfrac{(n+1)(5n+2)}{2}$

2. $\dfrac{(n+1)(2n+1)(2n+3)}{3}$

3. $\dfrac{n+1}{6}\,[6a^2 + 6adn + 2d^2n^2 + d^2n]$

4. $\dfrac{n^2(n+1)^2}{4}$

EXERCISE 13.2

2. $\dfrac{n}{2}$ **3.** np **4.** $\dfrac{1}{108}$

EXERCISE 13.3

2. 10,626

3. (a) 3876 (b) 1001

4. 141

6. $\binom{k + j - 1}{j}$

EXERCISE 13.4

2. (i) $t(n + 1, r) = t(n, r - 1) + rt(n, r)$
(iii) $\lambda^{(4)} + 2\lambda^{(3)} + \lambda^{(2)}$

EXERCISE 13.5

1. (a) $\dfrac{q^3 - 1}{q - 1}$ (b) $\dfrac{q^3 - 1}{q - 1}$ (c) $\dfrac{q^4 - 1}{q - 1}$

(d) $\dfrac{q^7 - 1}{q - 1} \cdot \dfrac{q^6 - 1}{q^2 - 1} \cdot \dfrac{q^5 - 1}{q^3 - 1}$ (e) $\dfrac{q^{10} - 1}{q - 1} \cdot \dfrac{q^9 - 1}{q^2 - 1}$

8. (a) $q^2 + q + 1$ (b) $q^2 + q + 1$ (c) $q^3 + q^2 + q + 1$

(d) $q^{12} + q^{11} + 2q^{10} + 3q^9 + 4q^8 + 4q^7 + 5q^6 + 4q^5 + 4q^4$
$+ 3q^3 + 2q^2 + q + 1$

(e) $q^{16} + q^{15} + 2q^{14} + 2q^{13} + 3q^{12} + 3q^{11} + 4q^{10} + 4q^9$
$+ 5q^8 + 4q^7 + 4q^6 + 3q^5 + 3q^4 + 2q^3 + 2q^2 + q + 1$

Chapter 14

EXERCISE 14.1

9. 1, 2, 3, 7, 15, 22

12. (i) 121; 242; 292 (ii) 141 (iii) 141; 65

EXERCISE 14.2

1. $n - 2, n - 3$

2. $\dfrac{(2n)!}{n!(n + 1)!}$

3. $\dfrac{(2n-2)!}{n!\,(n-1)!}$ **4.** $\frac{1}{2}(n-1)(n-2)+\dbinom{n}{4}$

5. $\dbinom{n}{r}$

EXERCISE 14.3

1. (a) $\dfrac{3}{64}$ (b) 0 (c) $\dfrac{459}{512}$ (d) $\dfrac{29}{512}$

2. (a) $\dfrac{5}{36}$ (b) $\dfrac{65}{972}$ (c) $\dfrac{2042}{6^5}$ (d) $\dfrac{3802}{6^5}$

3. (a) $\dfrac{242}{5^5}$ (b) $\dfrac{51}{500}$

4. (a) $\dfrac{3024}{5^6}$ (b) $\dfrac{57713}{5^7}$ (c) $\dfrac{7776}{5^7}$ (d) $\dfrac{29628}{5^7}$ (e) $\dfrac{40721}{5^7}$ (f) 0

EXERCISE 14.4

1. $\dfrac{3!}{\lambda_1!\lambda_2!\lambda_3!\lambda_4!\lambda_5!}$ **3.** 20

Chapter 15

EXERCISE 15.3

1. 1716   **2.** $\dfrac{312}{323}$ **3.** 10946

4. 10946   **5.** 1001   **6.** 2880

7. 12600

EXERCISE 15.4

1. 39   **2.** kF_n **3.** $hF_{n-1} + kF_n$

4. (i) $H_n = H_{n-1} + 2H_{n-2}$

(ii) Where $F(x) = H_1 + H_2 x + H_3 x^2 + \cdots$,

$$F(x) = \frac{1}{1 - x - 2x^2}$$

(iii) $H_n = \frac{1}{3}(2^n - (-1)^n)$

5. (i) $H_n = H_{n-1} + pH_{n-2}$

(ii) $F(x) = \dfrac{1}{1 - x - px^2}$

(iii) $H_n = \dfrac{1}{\sqrt{4p+1}}\left[\left(\dfrac{1 + \sqrt{4p+1}}{2}\right)^n - \left(\dfrac{1 - \sqrt{4p+1}}{2}\right)^n\right]$

Chapter 16

EXERCISE 16.2

2. $L^1, L^2, L^4, L^5, L^7, L^8$.   No

EXERCISE 16.4

7. $\dfrac{tk}{b}$, $\dfrac{tk(k-1)}{b(b-1)}$

EXERCISE 16.5

6. 40, 13, 4

Index

A

Affine planes
  axioms, 189, 190
  coordinates, 189
  exercises, 189, 193
Arithmetic power series
  discussion, 209
  exercises, 211
  summation method, 209
Arrangements
  circular, 39
  definition, 38
  examples, 40
  exercises, 41
  number of r-arrangements, 38
  with repetitions, 39, 40
Associative laws, finite fields, 181
Averaging method
  examples, 137
  existence, 137

B

Balanced incomplete block designs
  blocks, 265
  definition, 265
  examples, 267
  exercises, 267
  Steiner triple system, 267
  varieties, 265
Binomial coefficients
  examples, 52
  exercises, 47, 54
  Gaussian, 218, 221
  identities, 52, 53
  maximum term, 51, 52
  Pascal's formula, 48
Binomial distribution
  definition, 213
  examples, 211
  exercises, 213
Binomial identities, exercises, 55

Binomial random walk, example, 235
Binomial theorem
    examples, 46
    exercises, 47
    formulation, 45
    maximum term, 45
Bipartite graph, definition, 259
Birkhoff's theorem
    chromatic polynomial, 18
    proof, 20
Birth and death process, examples, 241
Blocks, balanced incomplete block designs,
        265
Boundaries, maps, 18
Branches, trees, 13

                        C

Cayley, counting labeled trees, 15
Chessboard, grains of wheat caper, 7
Chromatic polynomials
    Birkhoff's theorem, 18
    complete graph, 97
    definition, 18
    empty graph, 97
    examples, 18, 19, 68, 98, 101–105, 169
    exercises, 21, 69, 106, 173
    graphs, 21, 97, 168
    inclusion–exclusion principle, 68
    n-gon, 170
    overlapping graphs, 169
    properties, 171
    recurrence relations, 100, 105
    trees, 21, 170
    unsolved problems, 172
Chromatic triangles
    exercises, 174
    Ramsey problem, 174
Coloring
    graphs, 167
    maps, exercises, 21, 167
    planar maps, 167
Combinations
    definition, 42
    examples, 42, 82, 119, 215
    exercises, 43, 83, 85, 216, 240
    symbols, 34
    with repetitions, 82
    without repetitions, 42

Commutative laws
    finite fields, 181
    Pappus' theorem, 194
Complement, sets, 60
Computer
    evaluating polynomials, 24
    exercises, 26, 56, 59, 70, 108, 229, 232,
        245, 256, 260, 263, 275
Conjectures, number of regions of plane, 10
Continued fractions, exercises, 86
Contradiction method
    examples, 138
    exercises, 139
    existence, 138
Convex polyhedron, definition, 155
Convex sets
    convex hull, 140
    convex polygons, 230
    definition, 140
    exercises, 142
    existence, 140
Coordinates
    affine planes, 189
    Desargues' theorem, 194
    examples, 191
    exercises, 193
    homogeneous, 193
    Pappus' theorem, 194
Counting hairs
    Dirichlet's drawer principle, 22
    Fibonacci sequences, 249
Counting hares
    discussion, 249
    exercises, 249
    Fibonacci sequence, 249

                        D

Data consistency
    examples, 206
    exercises, 207
Derangements
    definition, 70
    examples, 70
    exercises, 72
Derivatives, generating functions, 121
Desargues' theorem
    exercises, 195
    finite fields, 194
    statement, 193

Difference methods, operators, 90
Difference equations
    definition, 92
    examples, 92, 123, 239, 240
    exercises, 93, 115, 120, 124, 232, 241
    Fibonacci sequence, 94, 242
    generating functions, 113
    initial conditions, 242
    nonempty subsets of set, 2
    random walks, 29
    subsets of set, 3
Difference operator
    definition, 90
    evaluating polynomials, 25
    examples, 90
    exercises, 26
Difference sets
    definition, 268
    examples, 268–270
    exercises, 270
Differential equations
    examples, 122
    generating functions, 122
Dirichlet's drawer principle
    counting hairs, 22
    examples, 22, 136
    exercises, 23, 138
    formulation, 136
Distributive law, finite fields, 181
Distribution of objects into boxes
    examples, 214, 215
    exercises, 216
Division ring, Desargues' theorem, 194

E

Edges
    graphs, 20
    maps, 18
    trees, 13
Equivalence classes
    discussion, 152
    examples, 152, 153
    exercises, 154
Euler's formula
    examples, 156–159
    exercises, 160
    method of averaging, 137
    polyhedra, 156
Events, examples, 200

Examples, *see* specific entries
Exercises, *see* specific entries
Exhaustion
    examples, 134
    exercises, 135, 185
    nonexistence, 134
Existence
    averaging method, 137
    by construction, 129
    contradiction method, 138
    convex sets, 144
    examples, 129, 133, 134
    exercises, 146, 148, 167
    by exhaustion, 133
    maps, 158, 161, 165
    orthogonal Latin squares, 262
    projective planes, 188, 192
    regular polyhedra, 163
    tessellations, 147
    trees, 22
    by trial and error, 134

F

Factorials, Stirling's formula, 58
Fano plane
    examples, 186
    finite, 186
Fibonacci algorithm
    example, 253
    exercises, 256
    formulation, 255
Fibonacci sequences
    counting hares, 249
    difference equation, 94
    examples, 118, 243, 247
    exercises, 95, 120, 245, 246, 249
    general term, 95
    generating functions, 243
    maximum and minimum, 250
    Pascal's triangle, 246
    representations, 242
    sequences of plus and minus signs, 247
Finite fields
    associate laws, 181
    cardinality, 184
    characteristic, 139, 184, 195
    commutative laws, 181
    definition, 180

discussion, 180
distributive law, 181
examples, 131, 138, 182, 184
exercises, 132, 139, 185
Finite planes
    exercises, 189
    Fano plane, 186
Formal power series, examples, 3
Four-color problem
    discussion, 164
    trivalent maps, 165

G

Gaussian binomial coefficients
    binomial coefficients, 221
    definition, 218
    examples, 218–223
    exercises, 224
Generating functions
    combinatorial identities, 116
    derivatives, 121
    difference equations, 113
    differential equations, 122
    examples, 3, 109, 111, 113, 114, 116, 119,
        121–124, 215, 227, 228, 234
    exercises, 8, 112, 115, 118, 120, 124, 229
    exponential, 110
    Fibonacci sequences, 243
    random walks, 29, 233
    Taylor series, 112
    triangulations, 231
Genus, definition, 165
Grains of wheat caper
    exercises, 8
    induction, 7
    story, 7
Graphs
    bipartite, 259
    chromatic polynomials, 21, 97, 168
    coloring, 167
    edges, 20
    examples, 169
    exercises, 168
    labeled, 20
    overlapping, 169
    planar, 167
    vertices, 20
    wheel, 168

H

Heawood, coloring theorem, 166

I

Identities
    binomial coefficients, 52
    combinatorial, 116
    generating functions, 116
Identity operator, definition, 90
Inclusion–exclusion principle
    applications, 67
    calculus of sets, 60
    derangements, 71
    examples, 64, 66, 83
    exercises, 66, 69, 81
    linear equations, 68, 78
    number theory, 67
    positive sets, 64
    probability, 201
    symmetric properties, 79
Induction
    exercises, 8
    number of nonempty subsets of set, 2
    random walks, 28
    regions of plane, 10
    Tower of Hanoi puzzle, 5
Initial conditions
    Fibonacci equation, 242
    random walks, 29
Interval of uncertainty, definition, 251
Irreducible, quadratic, 183
Iteration
    examples, 171
    exercises, 89
    recurrence relations, 87

K

Kirkman schoolgirl problem
    exercises, 264
    formulation, 263

L

Labeled graphs, example, 20
Labeled maps, example, 20

Labeled trees
  Cayley, 15
  definition, 14
  enumeration, 13
  example, 14
  exercises, 17
Labeling
  figures, exercises, 135
  Sperner's lemma, 175
Latin squares
  definition, 130, 260
  examples, 131, 260
  exercises, 132, 263
  existence, 262
  nonexistence, 262
  orthogonal, 131, 261
Line at infinity, definition, 192
Linear equations
  bounded solutions, 79
  examples, 74–79, 215
  exercises, 70, 76, 80, 229
  inclusion–exclusion principle, 68, 78
  nonnegative solutions, 74
  positive solutions, 73
  solutions bounded below, 73, 76, 78

M

Magic squares
  construction, 272
  definition, 271
  examples, 271
  exercises, 272
Maps
  boundaries, 18.
  coloring, 21
  edges, 18
  Euler's formula, 156
  examples, 129, 157, 158, 159
  exercises, 132, 160, 167
  existence, 158, 161
  five-color theorem, 165
  four-color problem, 164
  Heawood theorem, 166
  labeled, 20
  on sphere, 21, 155
  on surface of genus, 166
  ordinary, 18
  planar, 164

  polyhedral, 157
  regular, 161
Matchings
  exercises, 260
  maximal, 259
Maximum and minimum
  examples, 250, 252
  Fibonacci sequences, 250
Mean position, random walks, 29
Mercator projection, exercise, 161
Monic polynomial, definition, 27
Multinomial theorem
  enumeration of labeled trees, 16
  exercises, 57
  positive integer, 57
Multiple element sets, example, 61
Multiplication principle
  examples, 33
  formulation, 33
  number of r-arrangements, 38, 40

N

Nodes, trees, 13
Non-desarguesian planes, definition, 192
Nonempty subsets of set, exercises, 8
Nonexistence
  exercises, 139, 146
  by exhaustion, 134
  orthogonal latin squares, 262
  partition of cube, 146
  tiling rectangle, 144
Number of regions of plane, example, 10
Number theory, inclusion–exclusion principle, 67

O

Operators
  definitions, 90
  exercises, 93

P

Pappus' theorem
  exercises, 195
  finite fields, 194
  statement, 194

Partitions of integers
  definition, 226
  examples, 227, 228
  exercises, 229
Pascal's formula
  binomial coefficients, 48
  derivation, 48
  examples, 48, 49
  exercises, 53, 241
Pascal's triangle
  construction, 49
  examples, 49, 51, 117
  exercises, 54, 246
  Fibonacci sequences, 245
Permutations
  definition, 35
  derangements, 70
  examples, 35, 36, 70, 130
  exercises, 37, 41, 72
  product, 35
  symbols, 34, 35
Planes
  affine, 190
  Fano, 186
  non-desarguesian, 192
  projective, 187
Platonic solids
  discussion, 163
  exercises, 163
Polygons
  convex, 230
  rooted, 230
Polyhedra
  discussion, 137
  Euler's formula, 156
  existence, 163
  maps, 155, 157
  Platonic solids, 163
Polynomials
  difference operator, 25
  evaluating, 23
  exercises, 26
  monic, 27
Positive sets
  attributes, 64
  definition, 62
  derangements, 71
  examples, 63, 64, 66
  exercises, 63, 66
  inclusion–exclusion principle, 64
  number theory, 67

Power series, formal, 110
Probability
  discussion, 199
  distributions
    definition, 212
    exercises, 213
    random walks, 233
  examples, 200–203, 235
  exercises, 77, 203, 213, 238, 248
  inclusion–exclusion principle, 201
Product, permutations, 35
Projective configurations
  characteristic, 195
  Desargues, 193
  Pappus, 194
Projective planes
  axioms, 187
  cardinality, 188
  examples, 191
  exercises, 189, 193
  existence, 188, 192

Q

Quadratics
  exercises, 185
  over finite field, 183
  irreducible, 183

R

r-arrangements, example, 40
r-subsets
  example, 48
  exercises, 54
Ramsey problem, chromatic triangles, 174
Random walks
  binomial random walk, 234
  difference equation, 29
  examples, 27, 29, 49, 233, 234, 236, 237
  exercises, 30, 54, 238
  generating functions, 29, 233
  initial conditions, 29
  mean position, 29
  one-dimension, 27
  probability distributions, 233
  recurrence relation, 29
Recurrence relations
  chromatic polynomials, 15, 100

examples, 84, 88
exercises, 8, 84, 89
grains of wheat caper, 7
iteration, 87
random walks, 28
regions of plane, 10
Stirling numbers, 217
subsets of set, 2
Regions of plane
example, 8
exercises, 12
formula, 10
induction, 11
recurrence relation, 10
summation method, 12
Relative complement, sets, 60
Remainder theorem
exercises, 26
formulation, 24
Renewal theory, examples, 240
Room squares
definition, 273
examples, 273
exercises, 274

S

Sample spaces, examples, 200
Sequences of plus and minus signs
examples, 246, 247
exercises, 248
Sets
complement, 60
examples, 61
exercises, 8
inclusion–exclusion principle, 60
multiple elements, 60
number of nonempty subsets, 1
of permutations, 35
of subsets, 2, 53
positive 63
rth order 62
relative complement, 60
Sperner's lemma
exercises, 179
labeling triangles, 175
Steiner triple systems
definition, 267
exercises, 267
Stereographic projection, exercise, 161

Stirling numbers
discussion, 217
exercises, 218
recurrence relation, 217
Stirling's formula
examples, 58
exercises, 59
factorials, 58
Subsets
exercises, 54
r-subsets, 42
Summation method
arithmetic power series, 209
exercises, 56, 96, 211
general, 96
regions of plane, 12
Summation of series, examples, 91
Symbols
combinations and permutations, 34
number of r-arrangements, 38
permutations, 35
r-subsets, 42
Systems of distinct representatives
definition, 258
examples, 258, 259
exercises, 259

T

Temple of Benares, Tower of Hanoi, 6
Tessellations
exercises, 148
existence, 147
homogeneous, 148
of plane, 147
regular, 147
Tiling rectangle
into congruent squares, 143
examples, 145
exercises, 146
into incongruent squares, 143
nonexistence, 144
Tower of Hanoi
exercises, 8
induction, 5
puzzle formulation, 4
Temple of Benares, 6
Transformation, difference equation, 3
Translation operator, definition, 90

Trees
  branches, 13
  chromatic polynomials, 21, 170
  definition, 13
  edges, 13
  exercises, 17, 168
  existence, 22
  labeled, 13
  nodes, 13
  points, 13
  vertices, 13
Triangulations
  exercises, 179, 232
  generating functions, 231
  polygons, 176, 230
Trivalent maps, four-color problem, 165

U

Ultimate sets
  data consistency, 206

  definition, 205
  discussion, 204
  examples, 205, 206
  exercises, 207
Unsolved problems
  chromatic polynomials, 172
  four-color problem, 164

V

Varieties, balanced incomplete block designs, 265
Vertices
  graphs, 20
  trees, 13

W

Waiting lines, examples, 240